T0250135

The Story of Industrial Engineering

The Rise from Shop-Floor Management to Modern Digital Engineering

Analytics and Control
Series Editor: Adedeji B. Badiru, Air Force Institute of Technology, Dayton, Ohio, USA

Mechanics of Project Management
Nuts and Bolts of Project Execution
Adedeji B. Badiru, S. Abidemi Badiru, and I. Adetokunboh Badiru

The Story of Industrial Engineering
The Rise from Shop-Floor Management to Modern Digital Engineering
Adedeji B. Badiru

The Story of Industrial Engineering

The Rise from Shop-Floor Management to Modern Digital Engineering

Adedeji B. Badiru

CRC Press
Taylor & Francis Group
Boca Raton London New York

CRC Press is an imprint of the
Taylor & Francis Group, an **informa** business

CRC Press
Taylor & Francis Group
6000 Broken Sound Parkway NW, Suite 300
Boca Raton, FL 33487-2742

© 2019 by Taylor & Francis Group, LLC
CRC Press is an imprint of Taylor & Francis Group, an Informa business

No claim to original U.S. Government works

Printed on acid-free paper

International Standard Book Number-13: 978-1-138-61674-5 (Hardback)

Library of Congress Cataloging-in-Publication Data

Names: Badiru, Adedeji Bodunde, 1952-
Title: The story of industrial engineering : the rise from shop-floor
management to modern digital engineering / authored by Adedeji B. Badiru.
Description: Boca Raton : Taylor & Francis, a CRC title, part of the Taylor &
Francis imprint, a member of the Taylor & Francis Group, the academic
division of T&F Informa, plc, [2018] | Series: Analytics and Control
Identifiers: LCCN 2018036646| ISBN 9781138616745 (hardback : alk. paper) |
ISBN 9780429461811 (ebook)
Subjects: LCSH: Industrial engineering--History.
Classification: LCC T55.6 .B34 2018 | DDC 609--dc23
LC record available at https://lccn.loc.gov/2018036646

Visit the Taylor & Francis Web site at
http://www.taylorandfrancis.com

and the CRC Press Web site at
http://www.crcpress.com

*To the memory of Gary E. Whitehouse,
who put me on the path of Industrial
Engineering stories and accomplishments*

Contents

Preface

THIS FOCUS TEXT USES a storytelling approach to present the rich history of industrial engineering and its multitude of applications. This text traces the profession from its early days to the present day of digital engineering and additive manufacturing. The potpourri of stories includes the emergence of union movements, early human factors, early practices of ergonomics, the days of efficiency experts, the legend of cheaper by the dozen, the movement of scientific management, the evolution of manufacturing, the applications of management engineering, early management principles and movement, the emergence of formal industrial engineering, the diversification of industrial engineering, the risk of the fragmentation of industrial engineering, the offshoot professions and specializations, the struggle to keep the core (center) of industrial engineering, the mitigation of the splinter areas of the profession, the shadow functions of industrial engineering, the sustaining of the profession under a common core, the move into additive manufacturing applications, the concept of general digital engineering, and so on.

The text takes a relaxed and interesting storytelling tone to engage the interest of readers. The anecdotes recounted are based both on published literature as well as the author's own direct education, experience, and practice of the profession. This is not a textbook. Rather, it is a book of stories used to highlight the versatility and applicability of industrial engineering.

Acknowledgments

I EXPRESS PROFOUND GRATITUDE TO several colleagues, former teachers, and former students of industrial engineering, through whose insights I learned so much to create the platform for writing this book. My particular special thanks go to Dr. Sid G. Gilbreath, who has continued to mentor and inspire me over the years. His contributions to this book are found in the various classical reference materials that he provided to me as foundational stories for the manuscript. He provided leads and links to the cool stories of industrial engineering. I also acknowledge and appreciate the friendship, mentoring, and inputs of Dr. Bob Braswell, who provided the encouragement to write this book as a documentation to advance the future of industrial engineering.

Author

Adedeji B. Badiru is the dean and senior academic officer for the Graduate School of Engineering and Management at the Air Force Institute of Technology (AFIT), Dayton, Ohio. He holds a BS degree in industrial engineering, an MS degree in mathematics, an MS degree in industrial engineering from Tennessee Technological University, and a PhD in industrial engineering from the University of Central Florida. He is responsible for planning, directing, and controlling all operations related to granting doctoral and master's degrees, professional continuing cyber education, and research and development programs. Badiru was previously professor and head of systems engineering and management at AFIT; professor and department head of industrial and information engineering at the University of Tennessee, Knoxville, Tennessee; and professor of industrial engineering and dean of University College at the University of Oklahoma, Norman, Oklahoma. He is a registered professional engineer, a certified project management professional, a fellow of the Institute of Industrial Engineers, and a fellow of the Nigerian Academy of Engineering. His areas of interest include mathematical modeling, systems efficiency analysis, and high-tech product development. He is the author of more than 30 books, 35 book chapters, 75 technical journal articles, and 115 conference proceedings and presentations. He also has published 30 magazine articles and 20 editorials and periodicals. He is a member of several professional associations and scholastic honor societies. Badiru has won several awards for his teaching, research, and professional accomplishments.

CHAPTER 1

Personal Background for Industrial Engineering

Think like an IE, act like an IE.

ADEDEJI BADIRU

INDUSTRIAL ENGINEERING OF TODAY is different from the industrial engineering of the past. The job shops of the past are now the digital shops of today. This speaks to the diversity and flexibility of the profession.

The main premise of writing this focus book is to recognize, celebrate, and promote the diversity and versatility of the unadulterated profession of industrial engineering. It is written from my own personal observation, education, experience, and direct practice. It is sort of an eyewitness account of the glory of industrial engineering. I envision that the storytelling approach of this focus book will inspire the study, practice, and spread of industrial engineering.

In order to understand my fervent advocacy for industrial engineering, you will need to understand the story of how and why I was attracted to the profession in the first place, many decades ago. It is a very interesting background story. My first exposure to an industrial environment started in December 1972 in Lagos, Nigeria. Upon completion of my secondary school education at the famed Saint Finbarr's College, Akoka, Yaba, Lagos, Nigeria, I was employed as a factory supervisor at Associated Industries Limited (AIL) at Apapa, a mainland suburb of Lagos. The company made mints (peppermint). I got the employment barely two weeks after graduation from the secondary school. So, I did not have much time to decompress from the rigidly controlled educational regimen of Saint Finbarr's College, which was run as a tight ship by the principal, Reverend Father Denis Slattery, an Irish Priest. Father Slattery was popular throughout Nigeria at that time for his uncompromising commitment to seeing the school excel in academics, discipline, and sports. Fortunately, these three requirements fit my own personal attributes. So, I was a star scholar at the school. The discipline received under Father Slattery is still manifested today in my commitment to self-discipline. My Finbarr's-sparked interest in sports, focused primarily on soccer, has continued to this day. My focus on academics at the school later transferred into the pursuit of advanced degrees. The fact that I easily and quickly secured employment at Associated Industries soon after leaving high school was due to the fact that I graduated from Saint Finbarr's College with a Grade I Distinction in the West African Examination Council (WAEC) general external examination, which was required for graduation from high school in Nigeria at that time.

It was at AIL that I had my first taste of what would later become my profession of industrial engineering. At that time, I was not even aware of what industrial engineering was. What I knew was that some sort of better management was needed for AIL's production operations. I was assigned to the sugar-milling department. My job was to supervise casual labor employees in the process of pouring 100-pound bags of sugar into a large drum for the milling operation, which was one of the first stages of the peppermint production process. The factory engineer was an expatriate, who would always get furiously mad at me whenever the laborers mistakenly poured wet sugar into the machine, thereby causing a clog in the machine, which then necessitated production line shutdown. All of my workers were illiterate casual laborers, who did not know what was going on except that they were required to heave 100-pound bags of sugar over their heads to pour into the giant rotating drum, very much like a cement mixer. The workers could not peek into the drum to see what was going on inside of it.

It happened that wetness in the middle of the large bags of sugar could not be easily detected until the overhead pouring of the sugar into the large drum had started. Once gravity had taken over the downward flow of the granulated sugar, there was no manual avenue of stopping the flow without making a giant mess of spilled sugar all over the floor, which was a worse occurrence due to the unsalvageable loss of the raw material, the "almighty" costly sugar.

Visual or manual wetness inspection of the bags could not provide *a priori* foolproof assessment of the deep interior of each bag of sugar. There was no winning on the part of the workers. It was like a lottery pour each time. So, my team was usually at a loggerhead with the engineer. I suspected that the source of wetness in the bags of sugar was due to the poor storage facilities and the crude inventory system. Sometimes, retrieving the bags from storage to the production floor was a matter of last-in first-out, which was convenient, but not optimal.

On one of his usual tirades after a clogged machine, the engineer threatened to "throw" me, as the responsible floor supervisor, out the window of the second floor of the factory building, should wet sugar be found in the machine again. It was right there and then that I decided to resign from AIL in March 1973. It was my pride and arrogance from my Saint Finbarr's College education that made me want to quit and not subject myself to any further insult from the engineer. Graduates of Saint Finbarr's College were highly regarded and respected throughout Lagos in those days. We all went around town with our heads held high with pride. After all, we were the pupils and products of the prestigious school started in 1956 by Reverend Father Denis Slattery. Father Slattery was iconic throughout Lagos and many parts of Nigeria due to his multifaceted involvement in a variety of public issues, including being a school proprietor, a soccer referee, a coach, and a newspaper editor. So, for someone to assuage our reputation in public was the ultimate insult, which I was not ready to accept. So, I quit! In the United States a few years later, I analogized that quitting act to the scenes from Dolly Parton's *9 to 5* movie.

After leaving AIL, subsequently, between April 1973 and December 1975, I worked as an accounts clerk at Union Trading Company; a graphic artist and clerical officer at the Lagos State Ministry of Education, Audio Visual Aids Section; and a bank clerk at the Central Bank of Nigeria. All through the almost three years of working elsewhere after AIL, I still felt the injustice meted out to the manual workers at AIL. Sometimes, I would feel guilty for not sticking it out at the plant so that I could continue to protect the workers from the factory engineer. I wondered what level of oppressive treatment the workers would continue to endure if their new supervisor could not protect them against the engineer like I tried to do. My experience and observations of the treatment of the low-level workers at AIL would later form my attraction to the profession of industrial engineering, through which I envisioned I would become a staunch advocate for industrial workers.

Based on my excellent high school examination result, employers always suspected that I would depart their employment in favor of pursuing further studies at a university. It was usually a matter of consternation for everyone to see me working with only a high school diploma instead of immediately going to a university. But, in fact, I had my plan of when and how I would move forward to some sort of university-level education. I had earlier met my future spouse, Iswat, while we both worked at the Lagos State Ministry of

Education in 1973–1974. It was while I was working at the Central Bank of Nigeria that I received several merit-based academic scholarships. One scholarship was for studying mechanical engineering in Germany. One scholarship was for studying fine arts at the University of Nigeria, Nsukka. One scholarship was for studying medicine in Nigeria. One scholarship was for studying industrial engineering in the United States. Although I did not fully understand what industrial engineering entailed at that time, I was still attracted to it because of the "industrial" component in the name and my lingering loathing of what I observed at AIL production lines. It was much later that I would understand the linkage of bad worker treatments to the emergence of unions for the purpose of protecting workers' rights.

Faced with the opportunities of several fully paid scholarships, I chose to accept the scholarship for the study of industrial engineering in the United States. In addition to satisfying my "industrial" urge, the thrill of traveling to the United States played a role in my choice of which scholarship to accept. With a full scholarship from the Federal Republic of Nigeria, I could have gone to any university of my choice in the United States, with the normal admission process, of course. In my search for a U.S. university, I looked at names that rang a bell for me along the lines of "technological," "technical," "Institute," "Tech," etc. I did not discriminate among schools, because I did not know better regarding what constituted a better school in the United States. To me at that time, all U.S. schools were equally appealing; after all, they were all located in the great United States. So, the discriminant in my decision was the technical-sounding name of the school. It was for this reason that I applied to Tennessee Technological University just as well as I applied to Massachusetts Institute of Technology, Georgia Institute of Technology, California Institute of Technology, and Virginia Polytechnic Institute. Although I got accepted to several prestigious universities, I chose Tennessee Technological University. In my naïve mind of those days, it had the best-sounding name, and it accorded me the fastest and most responsive communication in the back-and-forth snail-mail iterations of the admissions process of the mid-1970s.

EXERCISING IE THINKING

The opening quote at the beginning of this chapter typifies how I have practiced, taught, mentored, and lectured industrial engineering for decades. Industrial engineers are known for making things better. In effect, industrial engineers make products better. I pride myself on being a product advocate, always thinking of how to help manufacturers make their products better. Many a time, I have directly contacted manufacturers to give them constructive feedback on my direct experience with their products and how I believe the products could be improved.

I shudder when consumers complain about products without offering constructive product-improvement suggestions based on direct usage experience. Product quality is a two-way affair. Quality, as designed and manufactured by the producer, is one thing. Quality, as practiced and perceived by the consumer, is a different thing. When a disconnect happens, many times it is because the user does not follow the user's guide provided by the manufacturer.

No product is ever made perfect from the beginning. Good quality often evolves over time and over several iterations of incremental improvements. A product that a manufacturer attempts to make perfect at the first introduction may never reach the market. A case example occurred when I visited a local Wuse Market in Abuja, Nigeria in 1994 with the goal of buying locally-made products as a part of my proactive product assessment efforts for the purpose of identifying flaws, the feedback for which could be used by the producers to improve their products. The seller insisted that I should buy an imported brand because it offered better quality. She was baffled when I insisted that it was the local brand that I desired to buy, not because of lower cost, but because that was my preference.

My views and thinking about product reviews are shaped by my industrial engineering education as narrated in the chapters that follow.

CHAPTER **2**

The Journey into Industrial Engineering

So it was, that I proceeded from Lagos, Nigeria, to study industrial engineering at Tennessee Technological University in Cookeville, Tennessee, on a full academic scholarship of the Federal Government of Nigeria. My travel took me from Lagos International Airport to London, United Kingdom, via Nigeria Airways on December 29, 1975. After spending the night of December 30 at a government-arranged hotel, I proceeded to New York City on the December 31, 1975, via the now defunct British Caledonia Airways. Being the eve of the new year of 1976, it was a difficult arrival in New York City. The Nigerian consulate in New York, where new Nigerian scholarship students were supposed to check in upon arrival in the United States, was already closed for the holidays by the time a taxi got me to the office in Manhattan, and I had no place to go. There were two other scholarship students, Kunle B. Aderogba and Ekpeyong, arriving at the closed office at the same time. We faced the predicament together. We worked together to figure out our next options. We had little money. The consulate was supposed to give us our initial stipends and living allowances upon arrival. The office being closed, we were out of luck. I had only $30.00 in traveler's checks that I had obtained via foreign currency exchange in Lagos at a cost of 22.74 Nigerian Naira. That was in the era when the Nigerian Naira was stronger than the U.S. dollar. It is amazing how things have changed. The other two students did not fare much better dollar-wise. It was getting dark very quickly, and the temperature was dropping fast. It was December winter weather in New York, and we did not have adequate winter clothing. The arrival plan was for the consulate to give us a winter clothing allowance, with which we were to immediately procure winter clothing from New York street stores located around the consulate office. A closed office meant no clothing allowance, which meant no adequate protection from the winter elements. The consulate was also supposed to make arrangements for our domestic transportation to our respective destinations, where our schools were located. Without an outbound transportation arrangement, we were grounded and stranded in Manhattan on the eve of the New Year. In search of a cheap hotel, the payment for which we planned to pool our financial resources, we set about roaming the streets around the consulate office, with our luggage propped on our heads, in the typical African market traders' fashion. As divine intervention would have it, a Nigerian man, returning home from work, spotted us and immediately figured out that we were new Nigerian students, stranded by the holiday closure of the consulate office. He came over to us and inquired what was going on. We explained our predicament to him. He introduced himself as Okike Onuoha. He said he would take us to his home and host us until the consulate would reopen. We were so delighted by his act of kindness and generosity. So it was that we spent two days in his home in Brooklyn. His wife, named Barbara, a Jamaican, received us warmly. We had a nice, warm, and comfortable time with the family from December 31, 1975, to January 2, 1976, when he took us to the Nigerian consulate office to continue our sojourn. After the administrative processes at the consulate, we three new students bade farewell to one another and proceeded our separate ways.

Unfortunately, we lost touch thereafter, primarily due to our poor communication infrastructure of those days and our immediate focus on the academic pursuits ahead rather than connectivity with ad hoc travel companions. From New York City, I boarded the now defunct Braniff Airlines to Nashville, Tennessee, from where I took a Trailways bus to Cookeville, Tennessee. On the bus, I met another Nigerian student, who had already been studying at Tennessee Tech for three years. He was Igbo from the Eastern part of Nigeria. I was Yoruba from the Western part, but he embraced me just like a fellow Igbo. He briefed me on what to expect in the school and the town of Cookeville. That was a nice initial orientation to what was to come. I arrived in Cookeville on the cold, cold evening of January 2, 1976. The rest from that point is now all history, to be recounted through other avenues in the future.

All through the long and arduous journey, my mind kept flashing back to the industrial experience of Associated Industries Limited and how my impending industrial engineering education would be an avenue for me to return to Nigeria and help clean up the basal industrial practices to make the lives of workers better. But first, I had to find out what this industrial engineering discipline was all about. All I knew was that I was interested in "industrial," and this new discipline offered "industrial" in its name. So, it must be good and fitting for me. The questions that danced around in my head in those early days included the following:

What is industrial engineering?

What do industrial engineers do?

How many industrial engineers exist?

Where do industrial engineers live? Do they reside in industrial quarters or in town?

Do industrial engineers wear white or brown overalls?

How could I use industrial engineering to solve Nigeria's industrial problems?

I was completely clueless, but I was determined to find out. I figured that the industrial engineering curriculum at my new appropriately named school, Tennessee Technological University, would reveal everything to me about the industrial engineering profession. I will not fail in the pursuit of this profession, I confided in myself.

So began my journey into the profession of industrial engineering. All I have learned, done, and professed about industrial engineering over the past four plus decades is formed and based on the initial accounts recounted in this chapter. This, thus, provides the foundation for the rest of this focus book on industrial engineering.

I present snippets of industrial engineering stories to highlight the fundamental principles of industrial engineering in simple, condensed, interesting, and engaging doses. It is hoped that the personal story presented at the beginning of this text will inspire further looks and studies of the profession. This personal story is the reason I went into industrial engineering, and it is the reason why I have remained true to the profession for over four decades. Hopefully, each reader can link his or her own story to a sustainable affinity for the profession of industrial engineering.

CHAPTER 3

Why Industrial Engineering?

T HE DIVERSE NEED OF the society is the reason why industrial engineering is needed. This fact has been highlighted by various handbooks and articles on industrial engineering, including those by Emerson and Naehring (1988), Zandin (2001), Martin-Vega (2001), Salvendy (2001), Sink et al. (2001), Badiru and Thomas (2009), Badiru (2014a), and Shtub and Cohen (2016).

Industrial engineering, more than any other discipline, is concerned with striking the best balance in the trade-offs of time, cost, and performance. Industrial engineers find ways to help organizations achieve their performance goals at the best combination of investment and schedule requirements.

The premise of this text is to incite interest in industrial engineering through applicable narratives of how the tools and techniques of the discipline apply to a broad spectrum of applications in business, industry, and engineering. Industrial engineering is the profession dedicated to making systems function better together with less waste, better quality, and fewer resources to serve the needs of society more efficiently and more effectively.

From the dawn of history, humans have sought ways to be more efficient and more effective in various pursuits. Humans in prehistoric times survived through industrious ingenuity and activities that we can today recognize as the practice of industrial engineering. The profession has a proud heritage with a direct link that can be traced back to the *industrial revolution*. Although the informal practice of industrial engineering has been in existence for centuries, there was no formal coalescing of the profession under one identifiable name. Named or not named, humans must practice industrial engineering in order to achieve their goals. The work of Frederick Taylor in the early twentieth century was the first formal emergence of the profession that would later be formalized into industrial engineering. It has been referred to with different names and connotations. Scientific management was one of the original names used to describe what industrial engineers do.

Human history indicates that humans started out as nomad hunters and gatherers, drifting to wherever food could be found. About 12,000 years ago, humans learned to domesticate both plants and animals. This agricultural breakthrough allowed humans to become settlers, thereby spending less time wandering in search of food. More time was, thus, available for pursuing stable and innovative activities, which led to discoveries of better ways of planting and raising animals for food. That initial agricultural discovery eventually paved the way for the agricultural revolution. During the agricultural revolution, mechanical devices, techniques, and storage mechanisms were developed to aid the process of agriculture. These inventions made it possible for more food to be produced by fewer people. The abundance of food meant that more members of the community could spend that time for other pursuits rather than the customary labor-intensive agriculture. Naturally, these other pursuits involved the development and improvement of the tools of agriculture. The extra free time brought on by more efficient agriculture was, thus, used to bring about more technological improvements in agricultural implements. These more advanced agricultural tools led to even more efficient agriculture. The transformation from the digging stick to the metal hoe is a good example of the raw technological innovation of that time. With each technological advance, less time was required for agriculture, thereby permitting more time for further technological advancements. The advancements in agriculture slowly led to more stable settlement patterns. These patterns led to the emergence of towns and cities. With central settlements away from farmlands, there developed a need for transforming agricultural technology to domicile technology that would support the new organized community settlements. The transformed technology was later turned to other productive uses, which eventually led to the emergence of the industrial revolution. To this day, the entwined relationships between agriculture and industry can still be seen. That is industrial engineering (regardless of whatever other names it is called).

This focus book presents the story of industrial engineering using a storytelling approach. It illustrates the various branches of industrial engineering and the diverse applications of the discipline. The storytelling approach helps to teach and elaborate the fundamental principles of industrial engineering in a simple, condensed, interesting, and engaging format. Is envisioned that the text will inspire deeper dives into the various facets of the profession of industrial engineering through formal education, short courses, seminars, practical workshops, and application demonstrations.

Industry, the root of the profession's name, clearly explains what the profession is about, although many modern practitioners, teachers, and researchers attempt to de-concatenate industry from the profession's name. This is a shame in my view. Rather than shying away from the linkage to "industrial" functions, we should embrace, promote, celebrate, and popularize the name of *industrial engineer* to eliminate any doubt about the value of the profession. The dictionary defines industry generally as the ability to produce and deliver goods and services. The "industry" in industrial engineering can be viewed as the application of skills and cleverness to achieve work objectives. This relates to how human effort is harnessed innovatively to carry out work. Thus, any activity can be defined as "industry" because it generates a product—be it a service or a physical product. A systems view of industrial engineering encompasses all the details and aspects necessary for applying skills and cleverness to produce work efficiently. The academic curriculum of industrial engineering must change, evolve, and adapt to the changing systems environment of the profession, without resorting to changing the name of the profession.

It is widely recognized that the occupational discipline that has contributed the most to the development of modern society is *engineering*, through its various segments of focus. Engineers design and build infrastructures that sustain the society. These include roads, residential and commercial buildings, bridges, canals, tunnels, communication systems, healthcare facilities, schools, habitats, transportation systems, and factories. Across all of these, the industrial engineering process of systems integration facilitates the success of the infrastructures. In this sense, the scope of industrial and systems engineering steps through the levels of activity, task, job, project, program, process, system, enterprise, and society. On the note of systems integration, Badiru (2014b) introduced the DEJI systems integration model that can take any process through the stages of design, evaluation, justification, and integration. It is through the final stage of integration that an organization can achieve goals and objectives with an optimized allocation of limited resources. That is industrial engineering.

From the age of horse-drawn carriages and steam engines to the present age of intelligent automobiles and aircraft, the impacts of industrial engineering cannot be mistaken, even though the contributions may not be recognized in the context of a specific application. A consistent usage of the term *industrial engineering*, buttressed by tangible results of its applications, will propagate the name. The more haphazardly we dillydally into other names, the more we diminish the core recognition of industrial engineering.

It is essential to recognize the alliance between "industry" and industrial engineering as the core basis for the profession. The profession has gone off on too many different tangents over the years. Hence, it has witnessed the emergence of industrial engineering professionals who claim sole allegiance to some narrow line of practice, focus, or specialization rather than the core profession. Industry is the original basis of industrial engineering, and it should be preserved as the core focus, which should be supported by the different areas of specialization. While it is essential that we extend the tentacles of industrial engineering to other domains, it should be realized that overdivergence of practice will not sustain the profession. The continuing fragmentation of industrial engineering is a major reason to write a convincing story of the profession. A fragmented profession cannot survive for long. The incorporation of systems can help to bind everything together without changing the profession's name. There is room enough to embrace and delve into other industry-related pursuits under the single name of "industrial engineering." The versatility of industrial engineering composes any of the following branches:

- Systems engineering
- Human factors
- Project management
- Traditional manufacturing
- Additive manufacturing
- Digital engineering
- Simulation modeling
- Management processes
- Operations management
- Production management
- Quality control
- Facility design
- Construction

- Service engineering

- Hospital management

- Healthcare services

- Supply chain management

- Logistics design and management

- Queuing analysis

- Operational sciences

- Operations research

Industrial engineering is the umbrella discipline for a variety of functional specializations. Whatever the need of business, industry, government, or the military happens to be, industrial engineering has it embedded within its core principles and functions. This is why industrial engineers are aggressively sought after by employers. But when the name of the profession is fragmented into other names, the unique name of industrial engineering is lost in the shuffle. On a personal note, I have a daughter who is a chemical engineer, but her employer utilizes her for industrial engineering functions. The employer does not call her an "industrial engineer." I also have a son who is a mechanical engineer, but his employer utilizes him for industrial engineering functions. The employer does not call him an "industrial engineer." These two examples are typical of examples that are rampant in many organizations. The functions are needed and should always be linked to a unified name that can promote the profession everywhere, anytime.

Notable industrial developments that fall under the purview of the practice of industrial engineering range from the invention of the typewriter to the invention of the automobile. Writing is a basic means of communicating and preserving records. It is one of the most basic accomplishments of the society. The course of history might have taken a different path if early writing instruments had not been invented at the time they were. The initial drive to develop the typewriter was based on the need and search for more efficient and effective ways of communication. The emergence of the typewriter typifies how an industrial product might be developed through the techniques of industrial engineering. In this regard, we can consider the chronological history of the typewriter:

1714	Henry Mill obtained a British patent for a writing machine.
1833	Xavier Progin created a machine that used separate levers for each letter.
1843	American inventor Charles Grover Thurber developed a machine that moved paper horizontally to produce spacing between lines.
1873	E. Remington & Sons of Ilion, New York, manufacturers of rifles and sewing machines, developed a typewriter patented by Carlos Glidden, Samuel W. Soule, and Christopher Latham Sholes, who designed the modern keyboard. This class of typewriters wrote in only uppercase letters but contained most of the characters on the modern machines.
1912	Portable typewriters were first introduced.
1925	Electric typewriters became popular. This made typeface to be more uniform. International Business Machines Corporation (IBM) was a major distributor for this product.

In each case of product development, engineers demonstrate the ability to design, develop, manufacture, implement, and improve integrated systems that include people, materials, information, equipment, energy, and other resources. Thus, product development must include an in-depth understanding of appropriate analytical, computational, experimental, implementation, and management processes. That is industrial engineering.

Going further back in history, several developments helped form the foundation for what later became known as industrial engineering. In the United States, George Washington was said to have been fascinated by the design of farm implements on his farm in Mt. Vernon. He requested an English manufacturer send him a plow built to his specifications that included a mold on which to form new irons when old ones were worn out or needed repairs. This can be described as one of the early attempts to create a process of achieving a system of interchangeable parts. This is industrial engineering. Thomas Jefferson invented a wooden mold board that, when fastened to a plow, minimized the force required to pull the plow at various working depths. This is an example of early agricultural industry innovation. This is industrial engineering. Jefferson also invented a device that allowed a farmer to seed four rows at a time. In pursuit of higher productivity, he invented a horse-drawn threshing machine that did the work of 10 men. This is industrial engineering.

Meanwhile in Europe, the Industrial Revolution was occurring at a rapid pace. Productivity growth, through reductions in manpower, marked the technological innovations of 1769–1800 Europe. Sir Richard Arkwright developed a practical code of factory discipline. In their foundry, Matthew Boulton and James Watt developed a complete and integrated engineering plant to manufacture steam engines. They developed extensive methods of market research, forecasting, plant location planning, machine layout, workflow, machine operating standards, standardization of product components, worker training, division of labor, work study, and other creative approaches to increasing productivity. This is industrial engineering. Charles Babbage, who is credited with the first idea of a computer, documented ideas on scientific methods of managing industry in his book entitled *On the Economy of Machinery and Manufacturers*, which was first published in 1832. The book contained ideas on division of labor, paying less for less important tasks, organization charts, and labor relations. These were all forerunners of formal industrial engineering.

Back in the United States, several efforts emerged to form the future of the industrial engineering profession. Eli Whitney used mass production techniques to produce muskets for the U.S. Army. In 1798, Whitney developed the idea of having machines make each musket part so that it could be interchangeable with other similar parts. By 1850, the principle of interchangeable parts was widely adopted. It eventually became the basis for modern mass production for assembly lines. It is believed that Eli Whitney's principle of interchangeable parts contributed significantly to the Union victory during the U.S. Civil War. This is industrial engineering.

Management attempts to improve productivity prior to 1880 did not consider the human element as an intrinsic factor. However, from 1880 through the first quarter of the twentieth century, the works of Frederick W. Taylor, Frank and Lillian Gilbreth, and Henry L. Gantt created a long-lasting impact on productivity growth through consideration of the worker and his or her environment. This is industrial engineering.

Frederick Winslow Taylor (1856–1915) was born in the Germantown section of Philadelphia to a well-to-do family. At the age of 18, he entered the labor force, having abandoned his admission to Harvard University due to impaired vision. He became an apprentice machinist and patternmaker in a local machine shop. In 1878, when he was 22, he went to work at the Midvale Steel Works. The economy was in a depressed state at the time. Frederick was employed as a laborer. His superior intellect was very quickly recognized. He was soon advanced to the positions of time clerk, journeyman, lathe operator, gang boss, and foreman of the machine shop. By the age of 31, he was made chief engineer of the company. He attended night school and earned a degree in mechanical engineering in 1883 from Stevens Institute. As a work leader, Taylor faced the following common questions:

Which is the best way to do this job?
What should constitute a day's work?

These are still questions faced by the industrial engineers of today. Taylor set about the task of finding the proper method for doing a given piece of work, instructing the worker in following the method, maintaining standard conditions surrounding the work so that the task could be properly accomplished, and setting a definite time standard and payment of extra wages for doing the task as specified. Taylor later documented his industry management techniques in his book entitled *The Principles of Scientific Management*. This is, indeed, industrial engineering.

The work of Frank and Lillian Gilbreth coincided with the work of Frederick Taylor. In 1895, on his first day on the job as a bricklayer, Frank Gilbreth noticed that the worker assigned to teach him how to lay brick did his work three different ways. The bricklayer was insulted when Frank tried to tell him of his work inconsistencies—when training someone on the job, when performing the job himself, and when speeding up. Frank thought it was essential to find one best way to do work. Many of Frank Gilbreth's ideas were similar to Taylor's ideas. However, Gilbreth outlined procedures for analyzing each step of workflow. Gilbreth made it possible to apply science more precisely in the analysis and design of the workplace. Developing *therbligs*, which is a spelling variation of "Gilbreth," as elemental predetermined time units, Frank and Lillian Gilbreth were able to analyze the motions of a worker in performing most factory operations in a maximum of 18 steps. Working as a team, they developed techniques that later became known as work design, methods improvement, work simplification, value engineering, and optimization. These are all industrial engineering. Lillian (1878–1972) brought to the engineering profession the concern for human relations. The foundation for establishing the profession of industrial engineering was originated by Frederick Taylor and Frank and Lillian Gilbreth. They were the first industrial engineers.

Henry Gantt's work advanced the management movement from an industrial management perspective. He expanded the scope of managing industrial operations. His concepts emphasized the unique needs of the worker by recommending the following considerations for managing work:

1. Define his task, after a careful study.
2. Teach him how to do it.
3. Provide an incentive in terms of adequate pay or reduced hours.
4. Provide an incentive to surpass it.

These are all steps that can be found in the steps of the practice of industrial engineering. Henry Gantt's major contribution is the Gantt chart, which went beyond the works of Frederick Taylor or the Gilbreths. The Gantt chart related every activity in the plant to the factor of time. This was a revolutionary concept for the time. It led to better production planning control and better production control. This involved visualizing the plant as a whole, like one big system made up of interrelated subsystems. As can be seen from the various accounts, industry has undergone a hierarchical transformation over the past several decades. Industry has been transformed from one focus level to the next, ranging from efficiency of the 1960s to the present-day trend of digital engineering and cyber operations.

In pursuing the applications of industrial engineering, it is essential to make a distinction between the tools, techniques, models, and skills of the profession. *Tools* are the instruments, apparatus, and devices (usually visual or tangible) that are used for accomplishing an objective. *Techniques* are the means, guides, and processes for utilizing tools for accomplishing the objective. A simple and common example is the technique of using a hammer (a tool) to strike a nail to drive the nail into a wooden workpiece (objective). A *model* is a bounded series of steps, principles, or procedures for accomplishing a goal. A model applied to one problem can be replicated and reapplied to other similar problems, provided the boundaries of the model fit the scope of the problem at hand. *Skills* are the human-based processes of using tools, techniques, and models to solve a variety of problems. Very important within the skills set of an industrial engineer are interpersonal skills or soft skills. This human-centric attribute of industrial engineering is what sets it apart from other engineering fields. What follows is a chronological listing of major events that can be ascribed to the evolution and practice of industrial engineering (in whatever form) over the centuries. As can be seen, industrial engineering, under whatever name it might have been called, has been around for a long time. Wherever efficiency, effectiveness, and productivity are involved, the basis is industrial engineering.

1440	Venetian ships are reconditioned and refitted on an assembly line.
1474	The Venetian Senate passes the first patent law and other industrial laws.
1568	Jacques Besson publishes an illustrated book on iron machinery as replacement for wooden machines.
1622	William Oughtred invents the slide rule.
1722	Rene de Reaunur publishes the first handbook on iron technology.
1733	John Kay patents the flying shuttle for textile manufacture—a landmark in textile mass production.
1747	Jean Rodolphe Perronet establishes the first engineering school.
1765	Watt invents the separate condenser, which made the steam engine the power source.
1770	James Hargreaves patents his "Spinning Jenny." Jesse Ramsden devises a practical screw-cutting lathe.
1774	John Wilkinson builds the first horizontal boring machine.
1775	Richard Arkwright patents a mechanized mill in which raw cotton is worked into thread.

1776   James Watt builds the first successful steam engine, which became a practical power source.

1776   Adam Smith discusses the division of labor in *The Wealth of Nations.*

1785   Edmund Cartwright patents a power loom.

1793   Eli Whitney invents the "cotton gin" to separate cotton from its seeds.

1797   Robert Owen uses modern labor and personnel management techniques in a spinning plant in the New Lanark Mills in Manchester, England.

1798   Eli Whitney designs muskets with interchangeable parts.

1801   Joseph Marie Jacquard designs automatic control for pattern-weaving looms using punched cards.

1802   The "Health and Morals Apprentices Act" in Britain aims at improving standards for young factory workers. Marc Isambard Brunel, Samuel Benton, and Henry Maudsey designed an integrated series of 43 machines to mass produce pulley blocks for ships.

1818   The Institution of Civil Engineers was founded in Britain.

1824   The repeal of the Combination Act in Britain legalizes trade unions.

1829   Mathematician Charles Babbage designs an "analytical engine," a forerunner of the modern digital computer.

1831   Charles Babbage published *On the Economy of Machines and Manufacturers.*

1832   The Sadler Report exposes the exploitation of workers and the brutality practiced within factories.

1833   Factory law is enacted in the United Kingdom. The Factory Act regulates British children's working hours. A general trades union is formed in New York.

1835   Andrew Ure publishes *Philosophy of Manfacturers.* Samuel Morse invents the telegraph.

1845   Friederich Engels publishes *Condition of the Working Classes in England.*

1847   The Factory Act in Britain reduces the working hours of women and children to 10 hours per day.
       George Stephenson founds the Institution of Mechanical Engineers.

1856   Henry Bessemer revolutionizes the steel industry through a novel design for a converter.

1869   The transcontinental railroad is completed in the United States.

1871   British Trade Unions are legalized by an act of Parliament.

1876   Alexander Graham Bell invents a usable telephone.

1877   Thomas Edison invents the phonograph.

1878   Frederick W. Taylor joins Midvale Steel Company.

1880   The American Society of Mechanical Engineers (ASME) is organized.

1881   Frederick Taylor begins time study experiments.

1885   Frank B. Gilbreth begins motion study research.

1886   Henry R. Towne presents the paper, *The Engineer as Economist*.
       The American Federation of Labor (AFL) is organized.
       Vilfredo Pareto publishes *Course in Political Economy*.
       Charles M. Hall and Paul L. Herault independently invent an inexpensive method of making aluminum.

1888   Nikola Tesla invents the alternating current induction motor, enabling electricity to take over from steam as the main provider of power for industrial machines.
       Herman Hollerith invents the electric tabulator machine, the first successful data processing machine.

1890   The Sherman Anti-Trust Act is enacted in the United States.

1892   Gilbreth completes motion study of bricklaying.

1893   Taylor begins work as a consulting engineer.

1895   Taylor presents a paper to ASME entitled *A Piece-Rate System*.

1898   Taylor begins time study at Bethlehem Steel.
        Taylor and Maunsel White develop a process for
        heat-treating high-speed tool steels.
1899   Carl G. Barth invents a slide rule for calculating
        metal cutting speed as part of a Taylor system of
        management.
1901   American national standards are established.
        Yawata Steel begins operation in Japan.
1903   Taylor presents a paper to ASMS entitled *Shop
        Management*.
        H.L. Gantt develops the "Gantt chart."
        Hugo Diemers writes *Factory Organization and
        Administration*.
        The Ford Motor Company is established.
1904   Harrington Emerson implements Santa Fe Railroad
        improvement.
        Thorstein B. Veblen writes *The Theory of Business
        Enterprise*.
1906   Taylor establishes metal-cutting theory for machine tools.
        Vilfredo Pareto writes the *Manual of Political Economy*.
1907   Gilbreth uses time study for construction.
1908   The Ford Model T is built.
        Pennsylvania State College introduces the first
        university course in industrial engineering.
1911   Taylor published *The Principles of Scientific
        Management*.
        Gilbreth publishes *Motion Study*.
        Factory laws are enacted in Japan.
1912   Harrington Emerson publishes *The Twelve Principles of
        Efficiency*.
        Frank and Lillian Gilbreth present the concept of
        therbligs.
        Yokokawa translates into Japanese Taylor's *Shop
        Management* and *The Principles of Scientific
        Management*.

1913   Henry Ford establishes a plant at Highland Park, Michigan, which utilizes the principles of uniformity and interchangeability of parts, and of the moving assembly line by means of conveyor belt.

Hugo Munstenberg publishes *Psychology of Industrial Efficiency*.

1914   World War I begins.

Clarence B. Thompson edits *Scientific Management*, a collection of articles on Taylor's system of management.

1915   Taylor's system is used at Niigata Engineering's Kamata plant in Japan.

Robert Hoxie publishes *Scientific Management and Labour*.

1915   Lillian Gilbreth earns a PhD in psychology at Brown University.

1916   Lillian Gilbreth publishes *The Psychology of Management*.

The Taylor Society is established in the United States.

1917   The Gilbreths publish *Applied Motion Study*.

The Society of Industrial Engineers is formed in the United States.

1918   Mary P. Follet publishes *The New State: Group Organization, the Solution of Popular Government*.

1919   Henry L. Gantt publishes *Organization for Work*.

1920   Merrick Hathaway presents the paper: *Time Study as a Basis for Rate Setting*.

General Electric establishes divisional organization.

Karel Capek presents *Rossum's Universal Robots*.

Capek's play coined the word "robot."

1921   The Gilbreths introduce process-analysis symbols to ASME.

1922   Toyoda Sakiichi's automatic loom is developed.

Henry Ford published *My Life and Work*.

1924   The Gilbreths announce results of micromotion study using therbligs.

Elton Mayo conducts illumination experiments at Western Electric.

1926   Henry Ford publishes *Today and Tomorrow*.

1927   Elton Mayo and others begin a relay-assembly test room study at the Hawthorne plant.

1929   The Great Depression begins.

The International Scientific Management Conference is held in France.

1930   Hathaway writes *Machining and Standard Times*.

Allan H. Mogensen discusses 11 principles for work simplification in *Work Simplification*.

Henry Ford publishes *Moving Forward*.

1931   Walter Shewhart publishes *Economic Control of the Quality of Manufactured Product*.

1932   Aldous Huxley publishes *Brave New World*, the satire that prophesies a horrifying future ruled by industry.

1934   General Electric performs micromotion studies.

1936   The word *automation* is first used by D.S. Harder of General Motors. It is used to signify the use of transfer machines that carry parts automatically from one machine to the next, thereby linking the tools into an integrated production line.

Charlie Chaplin produces *Modern Times*, a film showing an assembly line worker driven insane by the routine and unrelenting pressure of his job.

1937   Ralph M. Barnes publishes *Motion and Time Study*.

1941   R.L. Morrow writes *Ratio Delay Study*, an article in the *Mechanical Engineering* journal.

Fritz J. Roethlisberger writes *Management and Morale*.

1943   An ASME work standardization committee publishes a glossary of industrial engineering terms.

1945   Marvin E. Mundel devises a "memo-motion" study, a
       form of work measurement using time-lapse
       photography.
       Joseph H. Quick devises a work factors (WF) method.
1945   At a technical meeting of the Japan Management
       Association, Shigeo Shingo presents a concept of
       production as a network of processes and operations
       and identifies lot delays as sources of delay between
       processes.
1946   The first all electronic digital computer ENIAC
       (Electronic Numerical Integrator and Computer) is built
       at the University of Pennsylvania.
       The first fully automatic system of assembly is applied
       at the Ford Motor Plant.
1947   American mathematician Norbert Wiener writes
       Cybernetics.
1948   H.B. Maynard and others introduce the methods time
       measurement (MTM) method.
       Larry T. Miles develops value analysis (VA) at General
       Electric.
       Shigeo Shingo announces a process-based machine
       layout.
       The American Institute of Industrial Engineers is
       formed.
1950   Marvin E. Mundel writes *Motion and Time Study,
       Improving Productivity*.
1951   Inductive statistical quality control is introduced to
       Japan from the United States.
1952   A role and sampling study of industrial engineering is
       conducted at ASME.
1953   B.F. Skinner writes *Science of Human Behaviour*.
1956   A new definition of industrial engineering is presented
       at the American Institute of Industrial Engineering
       Convention.

1957   Chris Argyris writes *Personality and Organization.*
        Herbert A. Simon writes *Organizations.*
        R.L. Morrow writes *Motion and Time Study.*
1957   Shigeo Shingo introduces a scientific thinking
        mechanism (STM) for improvements.
        The Treaty of Rome established the European
        Economic Community.
1960   Douglas M. McGregor writes *The Human Side of
        Enterprise.*
1961   Rensis Lickert writes *New Patterns of Management.*
1961   Shigeo Shingo devises ZQC (source inspection) and
        poka-yoke systems.
1961   Texas Instruments patents the silicon chip integrated
        circuit.
1963   H.B. Maynard writes *Industrial Engineering Handbook.*
        Gerald Nadler writes *Work Design.*
1964   Abraham Maslow writes *Motivation and Personality.*
1965   Transistors are fitted into miniaturized "integrated
        circuits."
1966   Frederick Hertzberg writes *Work and the Nature of Man.*
1968   Roethlisberger writes *Man in Organization*
        The U.S. Department of Defense presents *Principles
        and Applications of Value Engineering*
1969   Shigeo Shingo develops single-minute exchange of dies
        (SMED).
        Shigeo Shingo introduces pre-automation.
        Wickham Skinner writes "Manufacturing—Missing
        Link in Corporate Strategy" article in *Harvard Business
        Review.*
1971   Taiichi Ohno completes the Toyota production
        system.
1971   Intel Corporation develops the microprocessor chip.
1973   The first annual Systems Engineering Conference of
        AIIE is held.

1975	Shigeo Shingo extols NSP-SS (non-stock production) system.
	Joseph Orlicky writes *MRP: Material Requirements Planning*.
1976	IBM markets the first personal computer.
1980	Matsushita Electric uses the Mikuni method for washing machine production.
	Shigeo Shingo writes *Study of the Toyota Production System from an Industrial Engineering Viewpoint*.
1981	Oliver Wight writes *Manufacturing Resource Planning: MRP II*.
1982	Gavriel Salvendy writes *Handbook of Industrial Engineering*.
1984	Shigeo Shingo writes *A Revolution in Manufacturing: The SMED System*.
1989	Code Division Multiple Access (CDMA) for cellular communications is developed.
1990	The concept of total quality management (TQM) becomes widely used.
1995	The dot-com boom starts in earnest.
	The Netscape search engine is introduced.
	Peter Norvig and Stuart Norvig publish *Artificial Intelligence: A Modern Approach*, which later became the authoritative textbook on artificial intelligence.
2000	The turning point of the twenty-first century and the Y2K computer date scare take place.
2004	Facebook social networking is born.
	Skype takes over worldwide online communication.
2008	The National Academy of Engineering (NAE) publishes the 14 Grand Challenges for Engineering.
2009	Adedeji Badiru and Marlin Thomas publish the *Handbook of Military Industrial Engineering* to promote the application of industrial engineering in national defense strategies. The handbook won the 2010 book-of-the-year Award from the Institute of Industrial Engineers.

2014	Adedeji Badiru publishes the second edition of the *Handbook of Industrial and Systems Engineering.* Unmanned Aerial Vehicles (UAV aka Drones) emerge as practical for a variety of applications.
2016	Self-driving cars begin to widely appear.
2017	The Internet of Things (IOT) makes a big splash.
2018	Adedeji Badiru publishes *The Story of Industrial Engineering.*

REFERENCES

Badiru, A. B. (Ed.). 2014a, *Handbook of Industrial and Systems Engineering*, 2nd ed. Boca Raton, FL: Taylor and Francis Group.

Badiru, A. B. 2014b, Quality Insights: The DEJI® Model for Quality Design, Evaluation, Justification, and Integration. *International Journal of Quality Engineering and Technology*, Vol. 4, No. 4, pp. 369–378.

Badiru, A. B. and Thomas, M. U. (Eds.). 2009, *Handbook of Military Industrial Engineering*. Boca Raton, FL: Taylor and Francis Group.

Emerson, H. P. and Naehring, D. C. E. 1988, *Origins of Industrial Engineering: The Early Years of a Profession.* Norcross, GA: Industrial Engineering and Management Press, Institute of Industrial Engineers.

Martin-Vega, L. A. 2001, The Purpose and Evolution of Industrial Engineering, in Zandin, K. B. (Ed.), *Maynard's Industrial Engineering Handbook*, 5th ed. New York, NY: McGraw-Hill.

Salvendy, G. (Ed.). 2001, *Handbook of Industrial Engineering: Technology and Operations Management*, 3rd ed. New York, NY: John Wiley and Sons.

Shtub, A. and Cohen, Y. 2016, *Introduction to Industrial Engineering*, 2nd ed. Boca Raton, FL: Taylor and Francis Group.

Sink, D. S. Poirier, D. F. and Smith, G. L. 2001, Full Potential Utilization of Industrial and Systems Engineering in Organizations, in Salvendy, G. (Ed.), *Handbook of Industrial Engineering: Technology and Operations Management*, 3rd ed. New York, NY: John Wiley and Sons.

Zandin, K. B. (Ed.). 2001, *Maynard's Industrial Engineering Handbook*, 5th ed. New York, NY: McGraw-Hill.

CHAPTER 4

Industrial Engineering Education

Welcome to the story of industrial engineering. Those who have known me over the years are familiar with my commitment to the purity of the industrial engineering profession as epitomized in my frequent ode to the profession, which is echoed in the box that follows.

I am an industrial engineer, I do systems.
I am an industrial engineer, I do economic analysis.
I am an industrial engineer, I do human factors.
I am an industrial engineer, I do simulation.
I am an industrial engineer, I do engineering management.
I am an industrial engineer, I do manufacturing systems.
I am an industrial engineer, I do operations management.
I am an industrial engineer, I do operations research.
I am an industrial engineer, I do whatever is needed to optimize and advance organizational performance, efficiency, and effectiveness.
Heck, I am an industrial engineer; I do it all. There is no shame in that.
A rose by any other name is still a rose. An industrial engineer by any other name is still an industrial engineer.

Industrial engineering does it all under one single unifying name. The preceding ode celebrates the diversity and versatility of industrial engineering under whatever name any group wishes to use for the profession. Industrial engineering, in its diversity, is concerned with striking the best trade-off balance between time, cost, quality, and performance. Industrial engineers find ways to advance organizations through the optimal pursuit of goals. The industrial engineering curriculum prepares students to design, develop, implement, and improve integrated systems of people, materials, information, equipment, energy, and other resources. The applications of industrial engineering are not limited to industry. Industrial engineers work and thrive in virtually every sector of the economy, whether in government, business, industry, military, and education. Industrial engineers use integrated and systematic tools and techniques from the analytical, computational, and experimental realms.

Have you ever wondered about the following questions?

- How can a product be designed to fit people, rather than forcing people to accommodate the product?

- How can merchandise layouts be designed to maximize the profit of a retail store?

- How can hospitals improve patient care while lowering cost?

- How can paper companies manage their forests (paper-making raw material) to both increase profits and still ensure long-term availability of trees?

- How can the work environment be designed to enhance comfort and safety while increasing productivity?

- How can a fast-food restaurant know how many and which kinds of burgers to have ready for the lunch-break rush?

- How can new car designs be tested before a prototype is ever built?

- How can space exploration be coordinated to link both management and technical requirements?

- How can a multipronged military attack be organized to sustain the supply chains?

Industrial engineers, using a systems-thinking approach, help answer and solve all of these questions. Industrial engineering thrives on systems perspectives just as systems thrive on industrial engineering approaches. One cannot treat topics of industrial engineering effectively without recognizing systems perspectives, and vice versa. One generic definition of an industrial engineering states the following:

> *Industrial Engineer*—One who is concerned with the design, installation, and improvement of integrated systems of people, materials, information, equipment, and energy by drawing upon specialized knowledge and skills in the mathematical, physical, and social sciences, together with the principles and methods of engineering analysis and design to specify, predict, and evaluate the results to be obtained from such systems.

This definition generically embodies the various aspects of what an industrial engineer does. Whether a definition is official or not, the description of the profession is always evolving as new application opportunities evolve. There is no need to change the name of the profession. It suffices to make the definition adaptive enough to embrace and promote new and emerging areas of application.

Industrial engineering is very versatile, flexible, and diverse with a strong anchor on rigorous systems thinking.

Some of the major functions of industrial engineers involve the following:

- Design integrated systems of people, technology, process, and methods.

- Develop performance modeling, measurement, and evaluation for systems.

- Develop and maintain quality standards for industry and business.

- Apply production principles to pursue improvements in service organizations.

- Incorporate technology effectively into work processes.

- Develop cost mitigation, avoidance, or containment strategies.

- Improve overall productivity of integrated systems of people, materials, and processes.

- Recognize and incorporate factors affecting performance of a composite system.

- Plan, organize, schedule, and control production and service projects.

- Organize teams to improve efficiency and effectiveness of an organization.

- Install technology to facilitate workflow.

- Enhance information flow to facilitate smooth operations of systems.

- Coordinate materials and equipment for effective systems performance.

In actual practice, industrial engineering makes systems function better together with less waste, better quality, and fewer resources.

The goal of every organization is to eliminate waste. Thus, the preceding definition is aptly relevant for everyone. Industrial engineering (IE) can be described as the practical application of the combination of engineering fields together with the principles of scientific management. It is the engineering of work processes and the application of engineering methods, practices, and knowledge to production and service enterprises. IE places a strong emphasis on an understanding of workers and their needs in order to increase and improve production and service activities. IE activities and techniques include the following:

1. Designing jobs (determining the most economic way to perform work)

2. Setting performance standards and benchmarks for quality, quantity, and cost

3. Designing and installing facilities

THE LINK TO SYSTEMS THINKING

Systems engineering involves a recognition, appreciation, and integration of all aspects of an organization or a facility. A system is defined as a collection of interrelated elements working together in synergy to produce a composite output that is greater than the sum of the individual outputs of the components. A systems view of a process facilitates a comprehensive inclusion of all the factors involved in the process.

INDUSTRIAL ENGINEERING EDUCATION AT TENNESSEE TECH

Even today, I still have fond memories of my IE professors at Tennessee Technological University—Sid Gilbreath, Jack Turvaville, James R. Smith, and, later, Meenakshi Sundaram. Sid Gilbreath was (and still is) an engaging, sociable, funny, magnanimous, and accommodating individual. He is widely experienced and broadly skilled on many practical pursuits. To this day, he has continued to be my mentor, advisor, and supporter. In the days of my undergraduate studies at Tennessee Technological University, I marveled at the knowledge base, span of expertise, and professionalism of many of my engineering instructors. Being a new international student, I wondered how such a high concentration of marvelous militarily experienced engineers could be found in one small institution in a nonglamorous part of the nation. I later found out that this was not an isolated incident in one institution. It turned out that a large number of engineering professors in the 1960s and 1970s across the United States had served in the U.S. Navy or Army during World War II and other wars of that era. After the wars, through GI Bill programs, many transferred their military training, education, and expertise into lecturing at the college level. The positive impression I had of my Tennessee Tech engineering professors gave me early incentives to apply more forthright efforts to my engineering education and, subsequently, choose academia as my career path. The consequence is that the foundational knowledge acquired from the military engineers turned professors continues to serve my own students in the years that followed my own education. My conclusion is that the military directly and indirectly influenced the advancement of technical manpower in the United States. What the United States is enjoying today in terms of being a world leader is predicated on a foundation of consistent technical education over the years. For this reason, investment in industrial engineering education is essential not only to keep the military on the cutting edge of warfare technology, but also to positively impact the national landscape of education on a broad scale.

RELEVANCE TO NATIONAL ACADEMY OF ENGINEERING'S 14 GRAND CHALLENGES

Recognizing the urgent need to address global societal issues from a technical standpoint, in 2008, the National Academy of Engineering (NAE) published the "14 Grand Challenges for

Engineering." The challenges have global implications for everyone, not just the engineering professions. As such, solution strategies must embrace all disciplines. Industrial engineering, by virtue of its global presence and wider span of application in science, technology, engineering, and mathematics (STEM) education can provide the technical and human foundation for addressing many of the challenges. Industrial engineers of the future will need diverse skills to tackle the multitude of issues and factors involved in adequately and successfully addressing the challenges. Industrial engineers, in particular, are needed to provide the diverse array of technical expertise, discipline, and professionalism required. STEM education provides a sustainable opportunity for all engineers to impact the 14 grand challenges as follows:

1. Make solar energy economical

2. Provide energy from fusion

3. Develop carbon sequestration methods

4. Manage the nitrogen cycle

5. Provide access to clean water

6. Restore and improve urban infrastructure

7. Advance health informatics

8. Engineer better medicines

9. Reverse-engineer the brain

10. Prevent nuclear terror

11. Secure cyberspace

12. Enhance virtual reality

13. Advance personalized learning

14. Engineer the tools of scientific discovery

The NAE document on the 14 grand challenges suggests that governmental and institutional, political and economic, and personal and social barriers will repeatedly arise to impede the pursuit of solutions to problems. As they have throughout history, engineers will have to integrate their methods and solutions with the goals and desires of all society's members. Who is better capable of analyzing, synthesizing, and integrating than industrial engineers?

CHAPTER 5

The Early Pioneers of Industrial Engineering

S EVERAL PIONEERS IN THE 1930s, 1940s, 1950s, and 1960s blazed the trail in the practice of industrial engineering. They left a trail of accomplishments that set the foundation for the advancement of the profession. Ralph M. Barnes was reportedly the first person to receive a PhD in industrial engineering. He spent many years laying the groundwork for revolutionary developments in the education and practice of industrial engineering. The impact of World War II accelerated the pattern and practice of industrial engineering. The pioneers stepped forward in droves to contribute to the war efforts and postwar developments.

RALPH MOSSER BARNES

Ralph Barnes was born on October 17, 1900, in Clifton Mills, West Virginia. He died on November 5, 1984. He received his BS and MS degrees at West Virginia University and his PhD at Cornell University. He worked at the U.S. Window Glass Company as an assistant engineer on product development in 1923; at Bausch & Lomb Optical Co. during 1924–1925; at the Gleason Works during 1925–1926 as an industrial engineer; at the University of Illinois as an instructor; at the College of Commerce during 1926–1928; at Eastman Kodak Company as an industrial engineer during the summers of 1927–1930 and during 1934–1936; at Kodak Ltd. in summer 1937; as assistant professor of industrial engineering at University of Iowa, College of Engineering during 1928–1930, associate professor of industrial engineering from 1930 to 1934, and professor of industrial engineering from 1934 to 1949, as well as director of personnel from 1937 to 1949 and director of the management course during 1938–1949; and professor of engineering and production management at the University of California, Los Angeles, from 1949 to his retirement. During World War II, he was a consulting engineer for companies in the east and midwest; he served as a consultant for the Federation of Norwegian Industries in summer 1950. He was awarded the Gilbreth Medal in 1941 for his extraordinary contributions to the field of Industrial Engineering and Management and the Industrial Incentive Award in 1951. In 1969 he received the Frank and Lillian Gilbreth Industrial Engineering Award.

Ralph Barnes contributed immensely to the method study and time study. He was the researcher and practitioner who continued building on the classic Gilbreth technique and philosophy and proclaimed that time study and micro-motions study were evidently different analysis techniques. He was a notable personality in the general field of industrial engineering and management. While at the University of Iowa (1928–1949), he and his students conducted many experiments and data analyses that established foundational techniques of industrial engineering. He continued to do this work at the University of California, Los Angeles (UCLA) from July 1, 1949, until his retirement on July 1, 1968. He is recognized for combining and coding all known data on motion analysis at that time. He classified the Principles of the Motion Economy into two important categories: "The Use of the Human Body" (1–16) and "The Design of Tools and Equipment" (17–22). One of his notable works is the *Rules of Barnes*. In 1931, his work *Industrial Engineering and Management* was published. In 1937 his worldwide famous standard work *Motion and Time Study* was published; the second edition was published in 1940, the third edition (559 pages) in 1949, and the fourth edition (665 pages) in January 1958. In 1944 the *Work Methods Manual* (136 pages) was published.

In 1949 his work *Motion and Time Study Problems and Projects* (220 pages) was published; the second edition (232 pages) was published in 1961. In 1949 his work *Motion and Time Study Applications* was published; the second edition (188 pages) was published in 1953, the third edition (188 pages) in 1958, and the fourth edition (188 pages) in 1961. In 1950 his *Work Methods Training Manual*, third edition (337 pages), was published. In 1951 the *Work Measurement Manual*, fourth edition (297 pages), was published. In 1956 the *Work Sampling* (264 pages) was published; the second edition (283 pages) was published in 1957. Ralph Barnes left an indelible mark on the profession of industrial engineering. The story of industrial engineering cannot be complete without good coverage of his multitude of contributions.

MARVIN EVERETT MUNDEL

Marvin Mundel was probably the second person, after Ralph Barnes, to receive a PhD in industrial engineering. Marvin Everett Mundel, born April 20, 1916, was a major figure in the fields of industrial engineering and time and motion studies. He is known particularly for his consulting work, seminars, and teaching, as well as numerous publications based on his expertise in work management and productivity enhancement. He began his engineering career in 1936 with a BS degree in mechanical engineering from New York University (1936), followed by MS and PhD degrees in industrial engineering earned in 1938 and 1939, respectively, from the Iowa State University.

In the late 1930s and 1940s, work measurement studies were considered the state-of-the-art method for improving industrial production. Mundel continued and built on the achievements of pioneers Frank and Lillian Gilbreth during his teaching career at both Bradley and Purdue universities. He also conducted seminars at Marquette University Management Center and the University of Wisconsin's Extension Center in Milwaukee. In addition to his American teaching career, Mundel was a visiting professor at both the University of Birmingham in England and Keio University in Tokyo, Japan.

In 1952, Mundel started a consulting firm that aided corporations and governments in either work measurement consulting or, later in his career, industrial engineering consulting. His first clients were U.S. government agencies that wanted to gain control over lost revenue or manpower. His position from 1952 to 1953 at the Rock Island Arsenal in Illinois, as the first director of the Army Management Engineering Training Program, transformed management techniques in the Army.

After his employment at Rock Island, Mundel began a series of consultant roles with corporations eager to standardize labor practices and make production more efficient. From 1953 to 1963, Mundel conducted time and motion studies at various manufacturing companies and developed techniques to measure work units. His most important contribution to the field of time and motion study was the development of memo-motion, a stop-action filming technique used to determine time standards for work tasks.

Following his refinements of time and motion study, Mundel took his expertise to Japan where he offered his consultant services to various Japanese manufacturing firms during the 1960s. His interests evolved from time and motion studies to include work management and overall management organization consulting. During the 1960s and 1970s, Mundel also returned to government consulting in the United States with these new techniques, in offices such as the Bureau of the Budget and U.S. Department of Agriculture. This period marked an important evolution in Mundel's career, from time and motion study to work measurement and then to industrial consulting. Mundel was among the first consultants to export American management techniques to Japan, and, in his later career, to other Asian countries. He became an integral part of the Asian Productivity Organization, a group that helped developing Asian countries learn how to increase productivity. His seminars sought to provide corporations and governments with efficient management techniques so that Asia would become a strong economic center. Mundel was sensitive to cultural differences as well as varied methods of management and standards of productivity.

Mundel won the Gilbreth Award in 1982. He continued conducting seminars and writing books and articles well into the 1980s, until failing health prevented him from traveling. When Mundel died in 1996, he was well respected in the field of industrial engineering for his many contributions.

WILLIAM A. GOLOMSKI

Bill Golomski was an educator, industrial worker, consultant, and manager of industrial engineering enterprises. Although his primary expertise was in the quality control area, his work cut across all parts of the practice of industrial engineering. The author of more than 300 papers and 10 books, Golomski holds master's degrees from the Milwaukee School of Engineering, the University of Chicago, Marquette University, and Roosevelt University. Golomski was named an honorary member of the American Society for Quality (ASQ) in 1992 in recognition of his "significant and enduring contributions to ASQ, including his service as president and chair in 1966–1968. In addition to his work in the corporate world, he has been a leader in communicating quality concept, principles, and methodology at the university level. As a distinguished teacher and educator, he has furthered the understanding of those concepts on which the Society's existence is based." Prior to his retirement, Golomski was president of W.A. Golomski and Associates, a Chicago-based international technical and management consulting firm. He was also a senior lecturer in business policy and quality management with the Graduate School of Business of the University of Chicago.

Those whose works preceded the work of Ralph Barnes, Marvin Mundel, and Bill Golomski might not have been bona-fide industrial engineers, but their works mirrored what industrial engineering would later entail. Their storied contributions are accounted for in the following paragraphs.

FREDERICK WINSLOW TAYLOR

Frederick Winslow Taylor was born on March 20, 1856, as a son of a wealthy Quaker family in Germantown, Pennsylvania, and died (of pneumonia) on March 21, 1915, in Philadelphia at age 59. He was recognized as the father of Scientific Management and one of the first management consultants in the industry. Trained as a mechanical engineer, Taylor's work was primarily akin to the practice of industrial engineering. For this reason, some people regard him as the first industrial engineer.

Even as a child he was noticed because of his strong passion for experiments and very precise analyses, which sometimes were almost obsessive. Because of an eye disease, he had to cancel his study of law as planned by his father, and in 1874 he started a course as a mechanical instrument maker and machine operator at the department of waterways in Philadelphia. After completion of the study, he could not find work suited for him. He, therefore, went to work at the Midvale Steel Company. Protected by the extraordinary attention of the president of Midvale, his reputation rose very quickly from apprentice worker to the "gang boss" level, assistant foreman, foreman machine room, work planner, chief mechanics, head of drawing/design room, and eventually to chief engineer. In addition to his normal work, in 1883 he completed a correspondence self-study (highly unusual at that time) in mechanical engineering at the Stevens Institute of Technology. Because of this study and his knowledge, he became head engineer at Midvale in 1884. Already during his first exercises, his ambitious attempts for rationalization led to some conflicts with management. He left Midvale Steel in 1890 to take up an office as general director of installations and labor of paper mills at Manufacturing Investments Co. He left this company in 1893 and became management consultant at Bethlehem Steel, where he conducted studies in loading steel and in steel operations, together with Maunsel White, who became very known and famous later on. This led to the development of high-speed steel (HSS). For his invention of the Taylor-White process for treatment of modern high-speed tools, he was rewarded with the gold medal of the Paris Exhibition in 1900. In 1901 he was fired after a disagreement with top management. Since he started his investigation in methods in 1875 to find the most efficient way to carry out tasks, and surely since he developed his method study and stopwatch technique at Midvale since 1878, he committed himself further to this pursuit. At age 45, he slowed down, primarily because of health problems. In 1903, he published his standard work *Shop Management*, and in 1911 he published *The Principles of Science Management*. He sustained himself based on incomes from his industrial activities and by patents incomes from his HSS invention. The followers of his techniques continued to refine and extend his Taylor's Principles. First, Taylor divided labor into various different suboperations, which enabled him to describe and classify the operations precisely. In doing this, he was able to measure the output and production precisely. This was called the "Taylor system."

Second, he saw the importance of training and education and in finding the right man for the right job, which led to better employee morale and positive motivation. By introducing this, he recognized that the division and distribution of work up to that point in time almost always led to bad labor relations and bad working conditions. Third, it was very important to provide every worker with detailed instructions and to supervise the correct execution of an assigned task. Fourth, he emphasized the importance of a good balance between labor and management, by good procedures and detailed descriptions. For that reason, he divided "management" into a number of 12 (sub) aspects of management. These sublevels were not all necessarily needed and might be combined. His classification and description of labor and management resulted in very efficient and smooth work processes. His scientific management was popularly known as Taylorism or the Taylor System, which is based on precise studies and analyses of human beings. The goal is to determine the correct sequence of motions for every human activity.

HENRY LAURENCE GANTT

Henry Laurence Gantt was born on May 20, 1861, in Calvert County, Maryland, and died on November 23, 1919, in Montclair, New Jersey, at the age of 58. He was a mechanical engineer and consultant. He was a protégé and coworker of Taylor from 1887 to 1893. He designed a wage system where the worker was paid a guaranteed daily income and a bonus for higher performance. He developed the famous Gantt charts that management used to indicate and visualize the progress of production by plotting on a straight line the real number of hours against the planned number of hours.

FRANK BUNKER AND LILLIAN GILBRETH

Frank Bunker Gilbreth (born July 7, 1868, in Fairfield, Maine, and died June 14, 1924, in Montclair, New Jersey, aged almost 56 years) and his wife Lillian Evelyn Möller Gilbreth (born May 24, 1878, in Oakland, California, and died January 2, 1972, in Phoenix, Arizona, aged 93 years) are the founders of motion analysis. After Frank resigned from construction work in 1912, they were able to focus themselves fully and spent all of their time on scientific management. He died suddenly of heart failure while making a telephone call from the railway station, leaving his wife behind with 11 children; daughter Mary had already died in 1912. They met each other more or less by accident in 1903 (Frank was president of his own construction firm), were married in October 1904, and raised 12 children between 1905 and 1922: Anne, Mary, Ernestine, Martha, Frank Jr, William, Lillian, Fred, Daniel, John,

Robert, and Jane. In 1948, Ernestine and Frank Jr. would write the famous book *Cheaper by the Dozen*. In 1950 the film *Cheaper by the Dozen* was released based on the book. Frank Sr. started working when he was 16 years old as an assistant bricklayer, became contractor, and ended up in management engineering. Later on he even became, by occasion, lecturer at Purdue University. He discovered his vocation when, as a young contractor, he tried to find a way to make bricklaying faster and easier. He observed that every bricklayer had come up with his own method of working, and not even two workers followed the same method. Soon, he developed many improvements in construction work and in the organization and supply of materials of the work. He demonstrated that bricklaying could be increased from 125 bricks per hour to 350 bricks per hour, by the following strategies:

- Eliminating unnecessary movements

- Supplying the bricks closer to the worker

- Designing special movable and adjustable scaffolds

It did not take long before the progressive Gilbreth became boss of his own successful contracting firm with offices in New York, Boston, and London.

This resulted in cooperation with Lillian Möller, who later became his wife, who studied working habits of factory workers and administrative workers in all kinds of industry, to find ways to increase output and to make work lighter and easier. Frank and Lillian in 1912 founded a consulting company focused on management consulting, Gilbreth, Inc.

He observed and analyzed human motions and activities, movements and motions with film shots (little lamps on the wrists) with an accuracy of even 1/2000 seconds and made three-dimensional models of copper wire. He divided these in the smallest possible elements, often consisting of no more than a simple single (finger) movement. Those smallest possible elements were classified in a limited number of categories (18 categories: Search [Sh], Find [Fi], Select [St], Transport Empty [TE], Grasp [G], Pre-position [Pp], Transport Loaded [TL], Position [P], Assemble [A], Dis-Assemble [DA], Release [Rl], Hold [H], Use [U], Plan [P], Inspect [I], Rest for overcoming fatigue [R], Unavoidable Delay [UD], and Avoidable Delay [AD]), which he named *standard elements*. To write these standard elements down quickly, he came up with simple symbols to denote them, called *therbligs*. They also came up with the two-hand analysis method called the SIMO-chart (simultaneous motion).

Now the problem emerged to find for each standard element the factors of influence and to determine the degree of influence. His intention was to determine synthetically the necessary time for each arbitrary motion or work from these standard elements. The big advantage would be that, in contrast to the division of each work into many smaller, always different, parts, as in the Taylor system, now each work could be built up from ever the same standard elements. He wanted to achieve this by a worldwide international common labor effort. He taught his children a 10-finger typing system (devised by himself) that resulted in them winning many typing prizes. Frank was the first to propose to appoint a surgical nurse as a "caddy" to hand over the prescribed instruments in the proper position in which they should be used by the surgeon. For that purpose, he used his "Packet Principle," through which the nurses learned to prepare the instruments in the same order as they would be used by the surgeon. Also, he formulated standard techniques to train recruits to disassemble and reassemble their weapons fast, even in the dark or when blindfolded.

Because of the early death (heart failure) of Frank B. Gilbreth, he was not able to work out his collection of data on times for his therbligs to a sound predetermined elemental motion time system (PEMTS). He was invited to address the International Management Committee in Prague around 1923, but he died just before he could attend. In his place, Lillian took over and presented his speech. Frank was a member of the American Society of

Mechanical Engineers and of the Taylor Society, and a lecturer at Purdue University.

Lillian graduated from the University of California, Berkeley, as a psychologist with a major in English literature, a BA in 1900 and MA in 1903, and pursued her PhD at Brown University. She also was a member of the American Society of Mechanical Engineers and lectured at Purdue University. She introduced psychology to management studies. After Frank died, Lillian continued his work under his name. In 1948 Lillian was chosen "Woman of the Year." In 1962, Frank (posthumously) and Lillian received the Frank and Lillian Gilbreth Industrial Engineering Award, named after them. Among the people who continued the principles of the Gilbreths are Ralph M. Barnes, Anne Shaw, and Alan Mogenson.

CHAPTER 6

Process Improvement in Industrial Engineering

INDUSTRIAL ENGINEERING HAS TRANSFORMED from the traditional shop-floor analysis to the more digital environment of applications. Both qualitative and quantitative applications are available in many areas of need in management, education, general business, industry, and the military. Process improvement is a primary basis for the practice of industrial engineering.

According to Heminger (2014), over the past few decades, a process approach has come to dominate our view of how to conceptualize and organize work. Current approaches to management, such as business process reengineering (BPR) (Hammer and Champy, 1993), lean (Womack and Jones, 2003), and Six Sigma (Pande et al., 2000), are all based on this concept. It seems almost axiomatic today to assume that this is the correct way to understand organizational work. Yet, each of these approaches seems to say different things about processes. What do they have in common that supports using a process approach? And, what do their different approaches tell us about different types of problems with the management of organizational work? To answer these questions, it may help to take a historical look at how work has been done since before the industrial revolution up to today.

Prior to the industrial revolution, work was done largely by craftsmen, who underwent a process of becoming skilled in their trade by satisfying customers' wants and needs. Typically, they started as apprentices, where they learned the rudiments of their craft from beginning to end, moved on to become journeymen, then became craftsmen as they become more knowledgeable, and finally, reaching the pinnacle of their craft as master craftsmen. They grew both in knowledge of their craft and in understanding what their customers wanted. In such an arrangement, organizational complexity was low, with a few journeymen and apprentices working for a master craftsman. But, because work by craftsmen was slow and labor intensive, only a few of the very wealthiest people could have their needs for goods met. Most people did not have access to the goods that the few at the top of the economic ladder were able to get. There was a long-standing and persistent unmet demand for more goods.

This unmet demand, coupled with a growing technological capability, provided the foundations for the industrial revolution. Manufacturers developed what Adam Smith (1776) called the "division of labor," in which complex tasks were broken down into simple tasks, automated where possible, and supervisors/managers were put in place to see that the pieces came together as a finished product. As we moved further into the industrial revolution, we continued to increase our productivity and the complexity of our factories. With the huge backlog of unmet demand, there was a willing customer for most of what was made. But, as we did this, an important change was taking place in how we made things. Instead of having a master craftsman in charge who knew both how to make goods as well as what the customers wanted and needed, we had factory supervisors, who learned how to make the various parts of the manufactured goods come together. Attention and focus began to turn inward from the customers to the process of monitoring and supervising complex factory work.

Over time, our factories became larger and ever more complex. More and more management attention needed to be focused inward on the issues of managing this complexity to turn out ever higher quantities of goods. In the early years of the twentieth century, Alfred Sloan, at General Motors, did for management what the industrial revolution had done for labor. He broke management down into small pieces and assigned authority and responsibility tailored to those pieces. This allowed managers to focus on small segments of the larger organization, and to manage according to the authority and responsibility assigned. Through this method, General Motors was able to further advance productivity in the workplace. Peter Drucker (1993) credits this internal focus on improved productivity for the creation of the middle class over the past one hundred years. Again, because of the long-standing unmet demand, the operative concept was that if you could make it, you could sell it. The ability to turn out huge quantities of goods culminated in the vast quantities of goods created in the United States during and immediately following World War II. This was added to by manufacturers in other countries, which came back online after having their factories damaged or destroyed by the effects of the war. As they rebuilt and began producing again, they added to the total quantities of goods being produced.

Then, something happened that changed everything. Supply started to outstrip demand. It did not happen everywhere evenly, either geographically or by industry. But, in ever-increasing occurrences, factories found themselves supplying more than people were demanding. We had reached a tipping point. We went from a world where demand outpaced supply to a world where more and more, supply outpaced demand (Hammer and Champy, 1993). Not everything being made was going to sell—at least not for a profit. When supply outstrips demand, customers can choose. And, when customers can choose, they will choose. Suddenly, manufacturers were faced with what Hammer and Champy call the "3 Cs": customers, competition, and change (Hammer and Champy, 1993). Customers were choosing among competing products, in a world of constant technological change. To remain in business, it was now necessary to produce those products that customers will choose. This required knowing what customers wanted. But, management and the structure of organizations from the beginning of the industrial revolution had been largely focused inward, on raising productivity and making more goods for sale. Managerial structure, information flows, and decision points were largely designed to support the efficient manufacturing of more goods, not on tailoring productivity to the needs of choosy customers.

BUSINESS PROCESS REENGINEERING

A concept was needed that would help organizations focus on their customers and their customers' needs. A process view of work provided a path for refocusing organizational efforts on meeting customer needs and expectations. On one level, a process is simply a series of steps, taken in some order, to achieve some result. Hammer and Champy, however, provided an important distinction in their definition of a process. They defined it as "a collection of activities that takes one or more inputs and creates an output that is of value to the customer" (1993). By adding the customer to the definition, Hammer and Champy provided a focus back on the customer, where it had been prior to the industrial revolution. In their 1993 book, *Reengineering the Corporation: A Manifesto for Business Revolution*, Hammer and Champy advocated BPR, which they defined as "the fundamental rethinking and radical redesign of business processes to achieve dramatic improvements in critical, contemporary measures of performance." In that definition, they identified four words that they believed were critical to their understanding of reengineering. Those four words were *fundamental, radical, dramatic,* and *processes.* In the following editions of their book, which came out in 2001 and 2003, they revisited this definition and decided that the key word underlying all of their efforts was the word *process*. And, with process defined as "taking inputs, and turning them into outputs of value to a customer," customers and customers' values are the focus of their approach to reengineering.

Hammer and Champy viewed BPR as a means to rethink and redesign organizations to better satisfy their customers. BPR would entail challenging the assumption under which the organization had been operating, and to redesign around their core processes. They viewed the creative use of information technology as an enabler that would allow them to provide the information capabilities necessary to support their processes while minimizing their functional organizational structure.

LEAN

At roughly the same time that this was being written by Hammer and Champy, Toyota was experiencing increasing success and buyer satisfaction through its use of Lean, which is a process view of work focused on removing waste from the value stream. Womack and Jones (2003) identified the first of the Lean principles as value. And, they state, "Value can only be defined by the ultimate customer." So, once again, we see a management concept that leads organizations back to focus on their customers. Lean is all about identifying waste in a value stream (similar to Hammer and Champy's process) and removing that waste wherever possible. But, the identification of what is waste can only be determined by what contributes or does not contribute to value, and value can only be determined by the ultimate customer. So, once again, we have a management approach that refocuses organizational work on the customers and their values.

Lean focuses on five basic concepts: value, the value stream, flow, pull, and perfection. "Value," which is determined by the ultimate customer, and the "value stream" can be seen as similar to Hammer and Champy's "process," which focuses on adding value to its customers. "Flow" addresses the passage of items through the value stream, and it strives to maximize the flow of quality production. "Pull" is unique to Lean and is related to the "just-in-time" nature of current manufacturing. It strives to reduce in-process inventory that is often found in large manufacturing operations. "Perfection" is the goal that drives Lean. It is something to be sought after, but never to be achieved. Thus, perfection provides the impetus for constant process improvement.

SIX SIGMA

In statistical modeling of manufacturing processes, *sigma* refers to the number of defects per given number of items created. Six Sigma refers to a statistical expectation of 3.4 defects per million items. General Electric adopted this concept in the development of the Six Sigma management strategy in 1986. While statistical process control can be at the heart of a Six Sigma program, General Electric and others have broadened its use to include other types of error reduction as well. In essence, Six Sigma is a program focused on reducing errors and defects in an organization. While Six Sigma does not explicitly refer back to the customer for its source of creating quality, it does address the concept of reducing errors and variations in specifications. Specifications can be seen as coming from customer requirements; so again, the customer becomes key to success in a Six Sigma environment.

Six Sigma makes the assertion that quality is achieved through continuous efforts to reduce variation in process outputs. It is based on collecting and analyzing data, rather than depending on hunches or guesses as a basis for making decisions. It uses the steps define, measure, analyze, improve, and control (DMAIC) to improve existing processes. To create new processes, it uses the steps define, measure, analyze, design, and verify (DMADV). Unique to this process improvement methodology, Six Sigma uses a series of karate-like levels (yellow belts, green belts, black belts, and master black belts) to rate practitioners of the concepts in organizations. Many companies that use Six Sigma have been satisfied by the improvements that they have achieved. To the extent that output variability is an issue for quality, it appears that Six Sigma can be a useful path for improving quality.

SELECTING A METHODOLOGY

From the descriptions, it is clear that while each of these approaches uses a process perspective, each addresses different problem sets, and they suggest different remedies. BPR addresses the problem of getting a good process for the task at hand. It recognizes that many business processes over the years have been designed with an internal focus, and it uses a focus on the customer as a basis for redesigning processes that explicitly address what customers need and care about. This approach would make sense where organizational processes have become focused on internal management needs, or some other issues, rather than on the needs of the customer.

The Lean methodology came out of the automotive world and is focused on gaining efficiencies in manufacturing. Although it allows for redesigning brand new processes, its focus appears to be on working with an existing assembly line and finding ways to reduce its inefficiencies. This approach would make sense for organizations that have established processes/value streams where there is a goal to make those processes/value streams more efficient.

Six Sigma was developed from a perspective of statistical control of industrial processes. At its heart, it focuses on variability in processes and error rates in production and seeks to control and limit variability and errors where possible. It asserts that variability and errors cost a company money, and learning to reduce these will increase profits. Similar to both BPR and Lean, it is dependent on top-level support to make the changes that will provide its benefits.

Whichever of these methods is selected to provide a more effective and efficient approach to doing business, it may be important to remember the lessons of the history of work since the beginning of the industrial revolution. We started with craftsmen satisfying the needs of a small base of customers. We then learned to increase productivity to satisfy the unmet demand of a much larger customer base, but in organizations that were focused inward on issues of productivity, not outward toward the customers. Now that we have reached a tipping point where supply can overtake demand, we need to again pay attention to customer needs for our organizations to survive and prosper. One of the process views of work may provide the means to do that.

REFERENCES

Drucker, P. F. 1993, *The Post Capitalist Society*. New York, NY: Harper Collins.

Hammer, M. and Champy, J. 1993, 2001, 2003, *Reengineering the Corporation: A Manifesto for Business Revolution*. New York, NY: Harper Business.

Heminger, A. R. 2014, Industrial Revolution, Customers, and Process Improvement, in Badiru, A. B. (Ed.), *Handbook of Industrial and Systems Engineering*. Boca Raton, FL: Taylor and Francis Group.

Pande, P. S., Neuman, R. P. and Cavanaugh, R. R. 2000, *The Six Sigma Way: How GE, Motorola, and Other Top Companies Are Honing Their Performance*. New York, NY: McGraw-Hill.

Smith, A. orig. 1776, 2012, *The Wealth of Nations*. London: Simon & Brown.

Womack, J. P. and Jones, D. T. 2003, *Lean Thinking: Banish Waste and Create Wealth in Your Corporation*. New York, NY: Free Press.

CHAPTER 7

Performance Measurement in Industrial Engineering

W HAT IS NOT MEASURED cannot be improved. Even if we subscribe to and embrace the principles of continuous improvement, no improvement can be pursued if there is no credible performance measurement strategy. Coleman and Clark (2014) present the techniques and tools for strategic performance measurement (Barrett 1999b, Bourne et al. 2002, Coleman et al. 2004, Deming 1960, 1993, Kennerley and Neely 2002, Lawton 2002, Leedy and Ormrod 2001, Muckler and Seven 1992, Neely et al. 1999).

The focus of this chapter is strategic performance measurement, a key management system for performing the study (or check) function of Shewhart's plan-do-study-act (PDSA) cycle. Strategic performance measurement applies to a higher-level system of interest (unit of analysis) and a longer-term horizon than operational performance measurement. While the dividing line between these two types of performance measurement is not crystal clear, the following distinctions can be made:

- Strategic performance measurement applies to the organizational level, whether of a corporation, a business unit, a plant, or a department. Operational performance measurement applies to small groups or individuals, such as a work group, an assembly line, or a single employee.

- Strategic performance measurement is primarily concerned with performance that has medium- to long-term consequences; thus performance is measured and reported on a weekly, monthly, quarterly, or annual basis. More frequent, even daily, measurement and reporting may also be included, but only for the most important performance measures. Data may also be collected, daily or perhaps continually, but should be aggregated and reported weekly or monthly. Operational performance measurement focuses on immediate performance, with reporting on a continual, hourly, shift, or daily basis. Strategic performance measurement tends to measure performance on a periodic basis, while operational performance measurement tends to measure on a continual or even continuous basis.

- Strategic performance measurement is concerned with measuring the mission- or strategy-critical activities and results of an organization. These activities and results are keys to the organization's success, and their measurements are referred to as strategic performance measures, key performance indicators, or mission-driven metrics. These measurements can be classified into a few key performance dimensions, such as Drucker's (1954) nine key results areas, the Balanced Scorecard's four performance perspectives (Kaplan and Norton 1996), the Baldrige criteria's five business results items (Baldrige Performance Excellence Program, 2011), or Sink's (1985) seven performance criteria.

- Strategic performance measurement tends to measure aspects of performance impacting the entire organization, while operational performance may be focused on a single product or service (out of many). In an organization with only one product, strategic and operational measurement may be similar. In an organization with multiple products or services, strategic performance measurement is likely to aggregate performance data from multiple operational sources.

- Strategic performance measurement is a popular topic in the management, accounting, industrial engineering, human-resources management, information technology, statistics, and industrial and organizational psychology literature. Authors such as Bititci et al. (2012), Brown (1996, 2000), Busi and Bititci (2006), Kaplan and Norton (1992, 1996), Neely (1999), Thor (1998), and Wheeler (1993) have documented the need for and the challenges facing strategic performance measurement beyond traditional financial and accounting measures. Operational performance measurement has long been associated with pioneers such as Frederick Taylor, Frank and Lillian Gilbreth, Marvin Mundel, and others. Careful reading of their work often shows an appreciation for and some application to strategic performance measurement, yet they are remembered for their contributions to operational measurement.

- For the remainder of this chapter, strategic performance measurement will be referred to as performance measurement. The term *measurement* will be used to apply to both strategic and operational performance measurement.

Why is performance measurement important enough to warrant a chapter of its own? Andrew Neely (1999, p. 210) summarized the reasons for the current interest in performance measurement very well. His first reason is perhaps the most important for the industrial engineer: the "changing nature of work." As industrialized nations have seen their workforces shift to predominantly knowledge and service work, concerns have arisen about how to measure performance in these enterprises with less tangible products. Fierce competition and a history of measuring performance have facilitated steady productivity and quality improvement in the manufacturing sector in recent years. Productivity and quality improvement in the service sector have generally lagged that of the manufacturing sector. The shift to a knowledge- and service-dominated economy has led to increased interest in finding better ways to measure and then improve performance in these sectors. Other reasons for increased interest in performance measurement cited by Neely include increasing competition, specific improvement initiatives that require a strong measurement component (such as Six Sigma or business process reengineering), national and international awards (with their emphasis on results, information, and analysis), changing organizational roles (e.g., the introduction of the chief information officer or, more recently, the chief knowledge officer), changing external demands (by regulators and shareholders), and the power of information technology (enabling us to measure what was too expensive to measure or analyze in the past).

MEASUREMENT IN THE CONTEXT OF PLANNING

An effective measurement approach enables and aligns individual, group, and organizational plan-do-study-act spirals to assist people in learning and growth toward a common aim. The plan-do-study-act spiral permeates human endeavor. Everything people do involves (consciously or unconsciously) four simple steps: make a plan, do the plan, study the results, and act the same or differently in the future, based on what was learned. Plan-do is the priority-setting and implementation process. Study-act is the measurement and interpretation process. Study-act is different than, yet inseparable from, plan-do. Plan-do-study-act is a structured and extremely useful (though mechanistic) theory of organizational learning and growth. The essence of plan-do-study-act within an organization is feedback and learning for the people in the system. Measurement's highest purpose in the context of strategy is to raise group consciousness regarding some phenomenon in the organization or its environment, thereby enhancing the opportunity to make mindful choices to further organizational aims. A strategic plan-do-study-act cycle for an organization may be notionally described by asking four fundamental questions: (1) What experiences and results does the organization aim to create over some time horizon? (2) How will people know if or when those experiences and results are occurring? (3) What actions and behaviors are people in the organization committed to, to create those experiences and results? (4) How will people know if those actions and behaviors are occurring? Questions (1) and (3) are strategic planning questions, while (2) and (4) are strategic measurement questions. Answers to questions (1) and (2) generally take the form of desired outcomes: nouns and adjectives. Answers to questions (3) and (4) generally take the form of planned activities: verbs and adverbs. Senior leaders have an obligation to answer questions (1) and (2) to provide direction and communicate expectations for the organization. Senior leaders are a participatory resource to help others in the organization shape answers to questions (3) and (4). One very important (though limited) view of leadership is the leader as organizational hypothesis tester: "If people act and behave question (3) answers— as verified by question (4) indicators—then question (1) results—as measured by question (2) indicators—are more likely to occur." This implicit hypothesis testing links planning and measurement in the management process.

THE MEASUREMENT AND EVALUATION PROCESS

Measurement is a human procedure of using language, images, and numbers to codify feedback from the universe about individual, organizational, and societal effectiveness—the extent, size, capacity, characteristics, quality, amount, or quantity of objects or events. In an organizational setting, measurement is the codifying of observations into data that can be analyzed, portrayed as information, and evaluated to support the decision maker. The term *observation* is used broadly here and may include direct observation by a human, sensing by a machine, or document review. Document review may involve secondary measurement, relying on the recorded observations of another human or machine, or it may involve the direct measurement of some output or artifact contained in the documents. The act of measurement produces data ("evidence"), often but not always in quantified form. Quantitative data are often based on counts of observations (e.g., units, defects, or person-hours) or scaling of attributes (e.g., volume, weight, or speed). Qualitative data are often based on categorization of observations (e.g., poor/fair/good) or the confirmation (or not) of the presence of desired characteristics (e.g., yes/no or pass/fail). Such qualitative data are easily quantified by calculating the percentages in each category.

Measuring performance—both strategic performance and operational performance—is a process that produces a codified representation of the phenomenon being measured. Assuming it was measured properly, this codified representation is simply a fact. This fact may exist in the form of a number, chart, picture, or text, and is descriptive of the phenomena being observed (i.e., organizational performance) and the process used to produce the fact prior to evaluation. Evaluation is the interpretation and judgment of the output of the measurement process (i.e., the number, chart, picture, or text). Evaluation results in a determination of the desirability of the level or trend of performance observed, typically on the basis of a comparison or expectation. Too often, those who are developing new or enhanced performance indicators jump to evaluation before fully completing the measurement step. They base the suitability of an indicator not on how well it represents the phenomena of interest, but on how it will be evaluated by those receiving reports of this indicator. As industrial engineers, we must know when to separate measurement from evaluation.

Phase 1: The process begins by asking what should be measured. Management or other stakeholders are interested in some event, occurrence, or phenomenon. This interest may be driven by a need to check conformity, track improvement, develop expectations for planning, diagnose problems, or promote accomplishments. This phenomenon of interest is often described in terms of key performance areas or criteria, which represent the priorities associated with this phenomenon.

Phase 2: The phenomenon of interest is observed or sensed to measure each key performance area (KPA). One or more indicators may be measured to represent the KPA. Each indicator requires an operational definition (a defined procedure for how the observation will be converted into data). While the KPAs are "glittering generalities," the indicators are specific and reliable.

Phase 3: The output of the measurement procedure is data, which are then captured or recorded for further use. Capturing represents entering the data into the "system," whether a paper or an electronic system. This step includes ensuring that all the data generated are captured in a timely, consistent, and accurate manner. This often includes organizing or sorting the data (by time, place, person, product, etc.) to feed the analysis procedures.

Phase 4: Raw data are analyzed or processed to produce information. Manual calculations, spreadsheets, statistical software packages, and other tools are used to summarize and add value to the data. Summarizing often includes aggregating data across time or units. That is, individual values are captured and processed; then, totals or means are calculated for reporting.

Phase 5: The output of analyzing the data is information, portrayed in the format preferred by the user (manager).

That is, when the values of the indicators representing KPAs for a particular phenomenon are measured, the portrayal should provide context that helps the user understand the information (Wheeler, 1993). Too often, the analyst chooses a portrayal reflecting his or her own preference rather than the user's preference.

Phase 6: The last step of the measurement and evaluation process is to perceive and interpret the information. How the user perceives the information is often as much a function of portrayal as content. (See Tufte's [1997a,b] work for outstanding examples of the importance of portrayal.) Regardless of which requirement (checking, improvement, planning, diagnosis, or promotion) prompted measurement, it is the user's perception of the portrayed information that is used to evaluate the performance of the phenomenon of interest. Evaluation results in continued measurement and evaluation, redesign of how the phenomenon is measured, or discontinuation and perhaps a transfer of interest to another phenomenon (Coleman and Clark, 2001).

PURPOSES OF STRATEGIC PERFORMANCE MEASUREMENT

Effective measurement demands that everyone understand why the measurement system is being created and what is expected from it. Design questions that arise during measurement system development can often be answered by referring back to the purpose of the system. Equally important is identification of all the users of the measurement system. If the system is being created for control purposes, then the manager or management team exerting control is the primary user. If the system is being created to support improvement, then most or the entire unit being measured may be users. The users should be asked how they will use the measurement system. Specifically, what kinds of decisions do they intend to make based on the information they receive from the measurement system? What information (available now or not) do they feel they need to support those decisions? The effectiveness of performance measurement is often dependent on how well its purpose and its user set are defined. That is, when one is evaluating whether a particular indicator is a "good" performance measure, one must first ask who will use it and what the intended purpose of the indicator is. An indicator that is good for one purpose or one user may not be as effective for another. Alternatively, an indicator that is potentially good for two or more purposes may best be used for only one purpose at a time. The use of the same indicator for potentially competing purposes, even though it could meet either purpose under ideal conditions, may lead to distortion (tampering), reluctance to report performance, or unexpected consequences, such as a lack of cooperation among the units being measured. In organizations, performance is typically measured for one or more of the following purposes:

- Control

- Improvement

- Planning

- Diagnosis

- Promotion

Control

Measuring performance for control may be viewed as measuring to check that what is expected has in fact occurred. Typically, a manager uses control indicators to evaluate the performance of some part of the organization for which the manager is responsible, such as a plant or department. A higher-level manager may have multiple units to control and require separate indicators from each unit. A lower-level manager may use indicators to control the performance of the individuals who work directly for that manager. In either case, the individual or unit whose performance is being monitored and controlled reports performance "upline" to the manager. If another part of the organization has the measurement responsibility (e.g., accounting and finance, quality control, or internal audit), it reports the most recent value of the indicators to the manager. The manager then reviews the level of performance on these indicators to check if the expectations are being met. Depending on the results of the comparison of current performance to expectations, and the manager's personal preferences, the manager takes action (or not) to intervene with the unit for the purpose of changing future levels of performance. Too often, managers only provide substantial feedback to the unit being evaluated when performance does not meet expectations. Control can be better maintained and performance improved when managers also reinforce good performance by providing feedback on expectations that are being met.

Care should be taken to distinguish between using an indicator to control the performance of an organizational unit and using the same indicator to judge the performance of the individuals managing or working in that unit. Measures of performance needed by managers may include elements of performance not completely within the control of those managing and working in that unit. For example, an indicator of total revenue generated by a plant may reflect the effectiveness of ongoing sales efforts, unit pricing pressure in the market, or a temporary downturn in the economy. While taking action in response to any of these factors may be appropriate for the senior-level manager who checks this plant's performance, judging the performance of local managers at the plant level by total revenue could lead to an emphasis on "making the numbers" over focusing on the factors that the local managers do control. "Making the numbers" in this situation could lead to such potentially undesirable consequences as building to inventory or spending for overtime to meet increased production targets generated by lower sales prices. A good rule of thumb is to measure performance one level above the level of control over results to encourage strategic action and to avoid suboptimization. At the same time, judgment of the performance of individual managers should focus on the causes and effects they control within the context of overall organizational performance. It is leadership's job to assist these managers in dealing with the factors beyond their control that affect their unit's overall performance.

Improvement

Measuring performance for improvement is more internally focused than measuring for control. Measuring for improvement focuses on measuring the performance of the unit one is responsible for and obtaining information to establish current performance levels and trends. The emphasis here is less on evaluating something or someone's performance, and more on understanding current performance levels, understanding how performance is changing over time, examining the impact of managerial actions, and identifying opportunities for improving performance. Managers often measure a number of things for use by themselves and their subordinates. An astute manager will identify drivers of end-result performance (e.g., sales, profits, and customer warranty claims) and develop indicators that lead or predict eventual changes in these end results. Such leading indicators might include employee attitudes, customer satisfaction with service, compliance with quality-management systems, and percent product reworked. Sears found that changes in store-level financial results could be predicted by measuring improvements in employee attitudes toward their job and toward the company. This predicted employee behavior, which, in turn, influenced improvements in customer behavior (customer retention and referral to other customers), leading, finally, to increases in revenue and operating margin (Rucci et al., 1998).

Employees, supervisors, and managers should be encouraged to establish and maintain indicators that they can use as yardsticks to understand and improve the performance of their units, regardless of whether these indicators are needed for reporting upline. Simply measuring a key performance indicator and making it promptly visible for those who deliver this performance can lead to improvement with little additional action from management. This assumes that those who deliver this performance know the desired direction for improvement on this indicator and have the resources and discretion to take actions for improvement. It is leadership's job to make sure the people in the organization have the knowledge, resources, discretion, and direction to use performance information to make improvements.

Planning

Measuring for the purpose of planning has at least two functions: (1) increasing understanding of current capabilities and the setting of realistic targets (i.e., goals) for future performance; and (2) monitoring progress toward meeting existing plans. One could argue that these simply represent planning-centric versions of measuring for improvement and then measuring for control. The role of measuring performance as part of a planning effort is important enough to warrant a separate discussion.

Nearly all strategic management or strategic planning efforts begin with understanding the organization and its environment. This effort is referred to as internal and external strategic analysis (Thompson and Strickland, 2003), organizational systems analysis (Sink and Tuttle, 1989), or, in plain words, "preparing to plan." A key part of internal analysis is understanding current performance levels, including the current value of key performance indicators and their recent trends. This provides the baseline for future performance evaluations of the effectiveness of the planned strategy and its deployment. Also, the choice of key performance indicators tells the organization what is important and is a specific form of direction often more carefully followed than narrative statements of goals and vision. Understanding current performance and its relation to current processes and resources provides managers with a realistic view of what is possible without having to make substantial changes to the system. Thus, setting intelligent targets for future performance requires an understanding of how implementation of the plan will change processes and resources to enable achievement of these targets. A key part of the external analysis is obtaining relevant comparisons so that the competitiveness of current performance levels and future performance targets can be evaluated. To answer the question of how good a particular performance level is, one must ask "compared to what?" Current competitor performance provides an answer to this question, but it must be assumed that competitors are also planning for improved performance. Setting future performance targets must take this moving competitive benchmark into account. Even the projected performance of your best current competitor may be inadequate as a future performance target to beat. The strategic management literature is full of examples of corporations that did not see their new competition coming and were blindsided by new competitors playing by different rules with substitutes for the bread-and-butter products of these corporations (see Hamel, 2002; Hamel and Prahalad, 1996). As Drucker (1998) has pointed out, some of the most important information managers need comes from outside their organizations and even outside their industries. A challenge for performance measurement is to provide not only internal but also external performance information that provides competitive intelligence for making strategic decisions.

Most strategic management or strategic planning processes include a last or next to last step that serves to measure, evaluate, and take corrective action. Often, this step is expected to be occurring throughout the process, with the formal execution of the explicit step occurring after goals have been set, action plans have been deployed, and strategy implementation is underway. That is, periodic review of progress toward meeting goals is a regular part of a strategic management effort, and performance indicators can provide evidence of that progress. When the goal-setting process includes the identification of key performance indicators and future performance targets for each indicator, the decision of which indicators to review has largely been made. In cases where goals are perhaps more qualitative or include simple quantitative targets without an operationally defined performance indicator, the planning team must choose or develop a set of progress indicators for these periodic (e.g., monthly or quarterly) reviews. A rule of thumb for these cases, based on the work of Sink and Tuttle (1989), is to develop indicators that provide evidence of the effectiveness, efficiency, quality, and impact of progress on each goal. Each of these terms is defined earlier. Even when key performance indicators have been predetermined at the time of goal setting, additional "drill-down" indicators may be required to explain performance trends and illustrate perceived cause-and-effect relationships among managerial actions, environmental and competitor actions, and observed levels of performance on end-result indicators.

Once the indicators have been chosen or developed, the periodic reviews are much more than collecting data, reporting current performance levels, and comparing to plan. How these reviews are conducted has a major impact on the organization's approach and even success with strategic management. If the reviews overemphasize checking or making sure that the people responsible for each goal are making their numbers, then reviews run the risk of taking on a confrontational style and may lead to gaming, distortion, and hoarding of information. Reviews that focus on what can be learned from the performance information and sharing lessons, and even resources when needed, can lead to better goal setting, improved action plans for implementing strategies, and increased sharing of performance information that may indicate future trends, good or bad. The type of review chosen is likely to reflect the organization's culture and the leadership's preferences. While either style may be used to drive performance, the two styles differ in the types of initiatives and actions leadership must take outside of and between periodic reviews to support performance improvement.

Diagnosis

Measuring performance for diagnosis or screening (Thor, 1998) is similar to the drill down described for illustrating cause-and-effect relationships among controllable and noncontrollable factors and their impact on end results. When an undesired (or desirable but unexplainable) result on a key indicator is observed, exploring the recent history of related indicators may provide insight into the possible causes. Tools such as the cause-and-effect (fishbone) diagram (Goal/QPC, 1985; Ishikawa, 1985) or quality function deployment (Akao, 1990) are useful in identifying drill-down metrics, likely to be at the cause of the observed effect. Unlike the previous methods, which are used for continual measurement of performance, measuring for diagnosis may be a one-time measurement activity with a start and an end. Thus, devoting resources to systematizing or institutionalizing the new indicators required should be based on the likelihood that these indicators will be needed again in the near future. When assessing the indicators of an existing measurement system, look for indicators once needed for diagnosis that have outlived their usefulness; stopping those outdated indicators may free up resources needed to produce newly identified indicators.

Promotion

Measuring for promotion (an idea contributed by Joanne Alberto) is using performance indicators and historical data to illustrate the capabilities of an organization. The intent is to go beyond simple sales-pitch claims of cutting costs by X percent or producing product twice as fast as the leading competitor. Here, the manager is using verifiable performance information to show the quantity and quality of product or service the organization is capable of delivering. Not only does this performance information show what is currently possible, it also provides a potential client with evidence that the organization measures (and improves) its performance as part of its management process. Thus, the customer can worry less about having to continually check this provider's performance and can rely on the provider to manage its day-to-day performance. A caveat here is that it is important to balance the organization's need to protect proprietary performance information with the customer's need for evidence of competitive product and service delivery. Care should also be taken in supporting the validity of promotional performance information so that the claims of less scrupulous competitors, who may boast of better levels of performance but present poorly substantiated evidence, are discounted appropriately.

Once the manager or engineer has clarified why performance is being measured, the question of what to measure should be addressed. Organizational performance is multidimensional, and a single indicator rarely meets all of the needs of the intended purpose.

DIMENSIONS OF PERFORMANCE

This section describes a number of frameworks for organizing the multiple dimensions of organizational performance. Each framework is a useful tool for auditing an organization's collective set of indicators to identify potential gaps. The intent here is neither to advocate the adoption of a specific framework as the measurement categories for a given organization, nor to advocate that an organization has at least one indicator for every dimension of these frameworks. The astute management team must recognize that organizational performance is multidimensional and make sure their measurement system provides performance information on the dimensions key to the success of their organization.

Those interested in a philosophical discussion of performance dimensions and how to choose the appropriate unit of analysis should read Kizilos (1984), "Kratylus Automates His Urnworks." This thought-provoking article sometimes frustrates engineers and managers who are looking for a single "correct" answer to the question of what dimensions of performance should be measured. The article is written in the form of a play with only four characters and makes excellent material for a group discussion or exercise.

The Concept of Key Performance Areas

Key performance areas are the few vital categories or dimensions of performance for a specific organization. KPAs may or may not reflect a comprehensive view of performance, but they do represent those dimensions most critical to that organization's success. While the indicators used to report the performance of each KPA might change as strategy or the competitive environment changes, the KPAs are relatively constant.

Rather than simply adopting one of the performance dimensions frameworks described in this section, an organization's managers should familiarize themselves with the alternative frameworks and customize the dimensions of their organizational scoreboard to reflect their organization's KPAs. What is most important is that the measurement system provides managers with the information necessary to evaluate the organization's performance in all key areas (i.e., KPAs) as opposed to conforming to someone else's definition of balance.

The Balanced Scorecard

While it has long been recognized that organizational performance is multidimensional, the practice of measuring multiple performance dimensions was popularized by the introduction of Kaplan and Norton's (1992) Balanced Scorecard. At its core, the Balanced Scorecard recognizes that organizations cannot be effectively managed with financial measures alone. While necessary for survival, financial measures tend to be lagging indicators of results and are frequently difficult to link to managerial actions aimed at improving medium- to long-term performance. Compounding this shortcoming, financial measurement systems are typically designed to meet reporting requirements for publicly traded companies or auditor's requirements for government agencies and privately held companies (i.e., financial accounting). Providing information to support management of the organization (i.e., managerial accounting) is an afterthought. This creates a situation where indicators developed for one purpose (fiscal control) are reused for another purpose (management and improvement), creating predictable problems.

The Balanced Scorecard views organizational performance from four perspectives, with the financial perspective being one of those four. The other three perspectives are the customer perspective, the internal process perspective, and the learning and growth perspective. Kaplan and Norton (1996) later suggested a general causal structure among the four perspectives. Thus, managerial actions to improve learning and growth, both at the individual and organizational levels, should result in improved performance on indicators of internal process performance, assuming the learning and growth initiatives and indicators are aligned with the internal process objectives. Improved performance on internal process indicators should result in improved results of the customer perspective indicators, if the process indicators reflect performance that is ultimately important to customers. And finally, if the customer perspective indicators reflect customer behaviors likely to impact the organization, then it is reasonable to expect improved performance on these customer indicators to lead to improved financial performance. For example, an initiative aimed at improving the quality assurance skills of quality technicians and quality management skills of production line supervisors might be indicated by increased numbers of Certified Quality Technicians and Certified Quality Managers (learning and growth indicators). Assuming this initiative was aimed at closing a relevant gap in skills, the application of these skills could be expected to improve levels of internal process indicators such as percent scrap and shift the discovery of defects further upline in the value stream (potentially reducing average cycle time for good product produced). Improvements in results on these internal process indicators could lead to fewer customer warranty returns, translating into direct financial savings. Improved performance on other customer-perspective indicators, such as customer perceptions of quality and their likelihood to recommend the product to others, although less directly linked, may also be predictors of improved financial results, such as increased sales.

While popular, the Balanced Scorecard has received some valid criticism. Nørreklit (2003) argues that the Balanced Scorecard has generated attention on the basis of persuasive rhetoric rather than on convincing theory. Theoretical shortcomings include suggested cause-and-effect relationships based on logic rather than empirical evidence and use of a strategic management system without addressing key contextual elements of strategic management (e.g., monitoring key aspects of the dynamic external environment or employing a top-down control model for implementation that appears to ignore organizational realities). Pfeffer and Sutton (2000, p. 148) point out that the Balanced Scorecard is "great in theory" but identify a number of problems in its implementation and use: "The system is too complex, with too many measures; the system is often highly subjective in its actual implementation; and precise metrics often miss important elements of performance that are more difficult to quantify but that may be critical to organizational success over the long term."

The industrial engineer's challenge is to sort through these shortcomings and address them with a well-designed measurement system that aligns with other management systems and balances practical managerial needs with theoretical purity. Practical issues related to designing and implementing a measurement system were previously described.

Richard Barrett (1999a) proposed enhancing the Balanced Scorecard by expanding the customer perspective to include suppliers' perspectives and adding three additional perspectives: corporate culture, community contribution, and society contribution. Certainly the importance of supply-chain management and partnering with suppliers warrants the inclusion of a supplier perspective in an organizational scorecard. Corporate culture has long been recognized as important to organizational success (Deal and Kennedy, 1982; Peters and Waterman, 1982) and appears as a key factor in the popular press accounts of great organizations. However, much work remains regarding how best to measure corporate culture and to use this information to better manage the organization. Management scholar Ralph Kilmann (1989; Kilmann and Saxton, 1983) and industrial engineer Larry Mallak (Mallak et al., 1997; Mallak and Kurstedt, 1996) offer approaches to measuring corporate culture. Off-the-shelf survey instruments, such as Human Synergistic's Organizational Culture Inventory, are also available. Barrett's recommendation to measure community and societal contributions are similar dimensions measured at different levels. Community contribution includes not only the cities, counties, and states where the organization and its employees reside and sell their products, but also the industries and professions in which the organization operates. Societal contribution expands beyond local impact and measures the organization's immediate and longer-term global impact.

The industrial engineer should recognize that the Balanced Scorecard is only one framework for organizing the dimensions of organizational performance and should be familiar with various alternatives to develop or adapt a framework that fits the organization's needs.

The Baldrige Criteria

A widely accepted performance dimensions framework that is updated biannually is the Results category of the Baldrige Criteria for Performance Excellence (Baldrige Performance Excellence Program, 2011). This category consists of five aspects that may be thought of as performance dimensions: product and process outcomes, customer-focused outcomes, workforce-focused outcomes, leadership and governance outcomes, and financial and market outcomes. When identifying indicators for each dimension, the Baldrige criteria stress choosing indicators that are linked to organizational priorities, such as strategic objectives and key customer requirements. The criteria emphasize segmenting results to support meaningful analysis and providing comparative data to facilitate the evaluation of levels and trends. The Baldrige criteria also include relative weights for each of these dimensions. The 2011–2012 version weights the last three dimensions (items) equally at 80 out of 450 total points. The product and process outcomes dimension is weighted slightly higher with 120 and the customer-focused outcomes dimension with 90 out of the total 450 points. Indicators of product and process outcomes provide evidence of the performance of products and processes important to customers. In the food service industry where customers want healthy eating alternatives, this might include providing comparisons of nutritional information of your leading products to those of key competitors. Process outcomes also include process effectiveness results for strategy and operations. Indicators of customer-focused outcomes provide evidence of the attitudes and behaviors of customers toward a company's products and services. This requires not only indicators of customer satisfaction and dissatisfaction, but also indicators of customer engagement, such as their willingness to recommend the company's products to others. Workforce-focused outcomes are particularly relevant to industrial engineers, because they include indicators of workforce capability and capacity and workforce engagement. Industrial engineers address the organization and management of work, including how work and jobs are organized and managed to create and maintain "a productive, caring, engaging, and learning environment for all members of your workforce" (Baldrige Performance Excellence Program, 2011, p. 48). Measuring the levels and trends of workforce capability and capacity could be an indicator of the performance of the industrial engineering function. Other items to be reported under workforce-focused outcomes include indicators of workforce climate, such as safety and absenteeism, workforce engagement such as turnover and satisfaction, workforce and leader development such as number of certifications and promotions. Such indicators are not just the domain of the human-resource manager, but include indicators that reflect the effectiveness of the work systems and supporting aids developed by the industrial engineers. The leadership and governance outcomes dimension starts with indicators of leadership communication and engagement to deploy vision and values and create a focus on action. Indicators providing evidence of effective governance and fiscal accountability might include financial statement issues and risks, and important auditor findings. This dimension also includes social responsibility results, addressing evidence of achieving and passing regulatory and legal requirements, indicators of ethical behavior and stakeholder trust, and indicators of the organization's support of its key communities. The final dimension in the Baldrige results framework is financial and market outcomes. This dimension includes traditional financial indicators such as return on investment and profitability and market indicators such as market share and market share growth.

Sink's Seven Criteria

D. Scott Sink provides an industrial engineer's view of performance with his seven performance criteria (Sink, 1985; Sink and Tuttle, 1989). He suggests that organizational performance can be described in terms of seven interrelated criteria:

- *Effectiveness*: Indicators of doing the correct things; a comparison of actual to planned outputs

- *Efficiency*: A resource-oriented criterion; a comparison of planned to actual resources used

- *Quality*: Defined by one or more of David Garvin's (1984) five definitions of quality (transcendent, product-based, manufacturing-based, user-based, or value-based) and measured at up to five (or six) points throughout the value stream

- *Productivity*: An indicator based on a ratio of outputs to the inputs required to produce those outputs (more on productivity later)

- *Innovation*: Indicators of organizational learning and growth as applied to the organization's current or future product and service offerings

- *Quality of work life*: Indicators of employee-centered results; preferably those predictive of higher levels of employee work performance

- *Profitability/budgetability*: Indicators of the relationship of revenues to expenses; whether the goal is to make a net profit or to stay within budget (while delivering expected levels of service)

Productivity

Productivity is a particularly important concept for industrial engineers and warrants further discussion here. Productivity indicators reflect the ratio of an organization's or individual's outputs to the inputs required to produce those outputs. The challenge is determining which outputs and inputs to include and how to consolidate them into a single numerator and denominator. Outputs include all the products and services an organization produces and may even include by-products. Inputs include labor, capital, materials, energy, and information.

Many commonly used productivity indicators are actually partial measures of productivity. That is, only part of the total inputs used to produce the outputs is included in the denominator. The most common are measures of labor productivity, where the indicator is a ratio of outputs produced to the labor inputs used to produce them (e.g., tons of coal per man day or pieces of mail handled per hour). While relatively simple and seemingly useful, care should be taken in interpreting and evaluating the results of partial productivity indicators. The concept of input substitution, such as increasing the use of capital (e.g., new equipment) or materials (e.g., buying finished components rather than raw materials), may cause labor productivity values to increase dramatically, owing to reasons other than more productive labor. A more recent shortcoming of measuring labor productivity is that direct labor has been steadily decreasing as a percent of total costs of many manufactured, mined, or grown products. In some cases, direct labor productivity today is at levels almost unimaginable 20 or 30 years ago. One might argue that the decades-long emphasis on measuring and managing labor productivity has succeeded, and that industrial engineers in these industries need to turn their attention to improving the productivity of materials and energy, and perhaps indirect labor. For more information, Sumanth (1998) provides a thoughtful summary of the limitations of partial productivity measures.

Total or multifactor productivity measurement approaches strive to address the limitations of partial productivity measures. Differing outputs are combined using a common scale such as constant value dollars to produce a single numerator, and a similar approach is used to combine inputs to produce a single denominator. Total factor approaches include all identifiable inputs, while multifactor approaches include two or more inputs, typically the inputs that make up the vast majority of total costs. The resulting ratio is compared with a baseline value to determine the percent change in productivity. Miller (1984) provides a relatively simple example using data available from most accounting systems to calculate the changes in profits due to any changes in productivity, as well as to separate out profit changes due to price recovery (i.e., net changes in selling prices of outputs relative to the changes in purchasing costs of inputs). Sink (1985) and Pineda (1996) describe multifactor models with additional analytical capabilities, useful for setting productivity targets based on budget targets and determining the relative contributions of specific inputs to any changes in overall productivity. Other approaches to productivity measurement such as data envelopment analysis (DEA) are beyond the scope of this chapter. See Cooper et al. (2004) and Medina-Borja et al. (2006) for further information about the use of DEA.

Quality

Quality, like productivity, deserves additional attention in an industrial engineer's view of measuring performance. Quality is ultimately determined by the end-user of the product or service.

And often, there are many intermediate customers who will judge and perhaps influence the quality of the product before it reaches the end-user. As there are numerous definitions of quality, it is important to know which definition your customers are using. While your first customer downstream (e.g., an original equipment manufacturer or a distributor) might use a manufacturing-based (i.e., conformance to requirements) indicator such as measuring physical dimensions to confirm they fall within a specified range, the end-user may use a user-based (i.e., fitness-for-use) indicator such as reliability (e.g., measuring mean time between failures [MTBF]) to evaluate quality. A full discussion of the five common definitions of quality and the eight dimensions of quality (performance, features, reliability, conformance, durability, serviceability, aesthetics, and perceived quality) is found in Garvin (1984). While seemingly adding confusion to the definition of quality within a larger performance construct, Garvin's eight dimensions of quality can be thought of as differing perspectives from which quality is viewed. Without multiple perspectives, one may get an incomplete view of a product's quality. As Garvin points out, "a product can be ranked high on one dimension while being low on another" (p. 30).

Once one or more definitions of quality have been chosen, the industrial engineer must decide where to measure quality prior to finalizing the indicators to be used. Sink and Tuttle (1989) describe quality as being measured and managed at five (later six) checkpoints. The five checkpoints correspond to key milestones in the value stream, with checkpoints two and four representing traditional incoming quality measurement (prior to or just as inputs enter the organization) and outgoing quality measurement (just before outputs leave the organization), respectively. Quality checkpoint three is an in-process quality measurement, a near-discipline in its own right, including statistical process control methods, metrology, certified quality technicians, and certified quality engineers. At checkpoint three we are measuring the key variables and attributes of processes, products, and services that predict or directly lead to the desired characteristics at outgoing quality measurement (quality checkpoint four) as well as those that contribute to success on the quality dimensions that are important further downstream (see checkpoint five). Tracking such variables and attributes lends itself to statistical analysis. See Chapters 3 and 9 for discussions about statistical process control. For an excellent introduction to applying statistical thinking and basic methods to management data, see Donald Wheeler's *Understanding Variation* (1993). The novice industrial engineer can benefit by taking heed of the late W. Edwards Deming's often stated admonition to begin by "plotting points" and utilizing the "most under-used tools" in management, a pencil and piece of grid paper. Quality checkpoint one is proactive management of suppliers and includes the indicators used to manage the supply chain. What might be incoming, in-process, outgoing, or overall quality management system indicators from the supplier's perspectives are quality checkpoint one indicators from the receiving organization's perspective. Quality checkpoint five is the measurement of product and service quality after it has left the organization's direct control and is in the hands of the customers. Quality checkpoint five might include indicators from the Baldrige items of product and service outcomes and customer-oriented results. Quality checkpoint five indicators provide evidence that products or services are achieving the outcomes desired by customers and the customer's reactions to those outcomes. The sixth, sometimes omitted, checkpoint is measuring the overall quality management or quality assurance process of the organization. Today we may relate this sixth checkpoint to the registration of an organization's quality management systems, as evidenced by receiving an ISO 9001 certificate.

Human Capital

Industrial engineers have long been involved in the measurement and evaluation of the performance of individuals and groups. As the knowledge content of work has increased, the overall cost and value of knowledge workers have increased. Organizations spend substantial energy and resources to hire, grow, and retain skilled and knowledgeable employees. Although these expenditures are likely to appear in the income statement as operating costs, they are arguably investments that generate human capital. While an organization does not own human capital, the collective knowledge, skills, and abilities of its employees represent an organizational asset—one that should be maintained or it can quickly lose value. Organizations need better measurement approaches and performance indicators to judge the relative value of alternative investments that can be made in human capital. They need to know which are the most effective options for hiring, growing, and keeping talent. The following paragraphs provide the industrial engineer with context and examples to help tailor their performance measurement toolkit to the unique challenges associated with measuring the return on investments in human capital.

Traditional human-resource approaches to measuring human capital have focused on operational indicators of the performance of the human-resources function. In particular, these indicators have emphasized the input or cost side of developing human capital. Such indicators might include average cost to hire, number of days to fill an empty position, or cost of particular employee benefits programs. More holistic approaches (Becker et al., 2001) focus on business results first, and then link indicators of how well human capital is being managed to create those results.

Assuming the organization has developed a multidimensional performance measurement system as described in this chapter, the next step is to identify human capital–related drivers of the leading organizational performance indicators (e.g., product and process outcomes, customer-focused outcomes as opposed to lagging performance results such as financial and market outcomes). Such drivers are likely to be related to employee attitudes and behaviors. Drivers of customer-focused outcomes might include employee attitudes toward their jobs or supervisors, or behaviors such as use of standard protocols and knowing when to escalate an issue to a customer-service manager. Drivers of product and process outcomes might include behaviors such as use of prescribed quality assurance procedures, completing customer orientation upon delivery, or perhaps an organizational effectiveness indicator such as cycle time (i.e., where cycle time is heavily dependent on employee performance). Indicators of the health of an organization's human capital are likely to predict or at least lead performance on these human capital–related drivers of organizational performance. Indicators of the health of human capital reflect the value of human capital as an organizational asset. Examples of such indicators include average years of education among knowledge workers (assumes a relatively large pool of employees), a depth chart for key competencies (i.e., how many employees are fully qualified to fulfill each mission), attrition rates, or more sophisticated turnover curves that plot turnover rates in key positions by years of seniority. Finally, traditional cost-oriented measures of human-resource programs can be evaluated in terms of their impact on the health of human capital and human capital drivers of organizational performance.

Human capital indicators should help answer questions such as the following: Does the new benefit program reduce turnover among engineers with 10–20 years of experience? Does the latest training initiative expand our depth chart in areas that were previously thin, thus reducing our risk of not being able to meet product and service commitments? Do changes to our performance management system improve employee attitudes among key customer interface employees? Do our initiatives aimed at improving employee attitudes and behaviors translate into better products and services as well as customers who increase the percentage of their business they give to our organization? Measuring human capital and the return on investments in human capital are new frontiers in measurement for industrial engineers, with the potential to make substantial contributions to organizational competitiveness.

IMPLEMENTING A MEASUREMENT SYSTEM

Once clear about why to measure performance and what dimensions of performance to measure, the question becomes how to implement a functioning measurement system. The measurement system includes not only the specific indicators, but also the plans and procedures for data gathering, data entry, data storage, data analysis, and information portrayal, reporting, and reviewing. A key recommendation is that those whose performance is being measured should have some involvement in developing the measurement system. The approaches that can be used to develop the measurement system include the following: (1) have internal or external experts develop it in consultation with those who will use the system; (2) have the management develop it for themselves and delegate implementation; (3) have the units being measured develop their own measurement systems and seek management's approval; or (4) use a collaborative approach involving the managers, the unit being measured, and subject matter expert assistance. This last approach can be accomplished by forming a small team—the measurement system design team.

A *design team* is a team whose task is to design and perhaps develop the measurement system; however, day-to-day operation of the measurement system should be assigned to a function or individual whose regular duties include measurement and reporting (i.e., it should be an obvious fit with their job and be seen as job enrichment rather than an add-on duty unrelated to their regular work). When ongoing performance measurement is assigned as an extra duty, it tends to lose focus and energy over time and falls into a state of neglect. Depending on how work responsibility is broken down in an organization, it may make sense to assign responsibility for measurement system operation to industrial engineering, accounting and finance, the chief information officer, quality management/assurance, human resources, or a combination of these. The design team should include the manager who "owns" the measurement system, a measurement expert (e.g., the industrial engineer), two or more employees representing the unit whose performance is being measured, and representatives from supporting functions such as accounting and information systems.

Each of the four development approaches can benefit from adopting a systems view of the organization using an input/output analysis.

Input/Output Analysis Using the SIPOC Model

A tool for helping users identify information needs at the organizational level is the input/output analysis or SIPOC (suppliers, inputs, processes, outputs, and customers) model. The intent is to get the users to describe their organization as an open system, recognizing that in reality there are many feedback loops within this system that make it at least a partially closed loop system. The SIPOC model is particularly useful for the design team approach to developing a measurement system. The model helps the team members gain a common understanding of the organization and provides a framework for discussing the role and appropriateness of candidate indicators.

The first step to complete the SIPOC model is to identify the organization's primary customers, where a customer is anyone who receives a product or service (including information) from the organization. Next identify the outputs, or specific products and services, provided to these customers: for an organization with a limited number of products and services, these can be identified on a customer-by-customer basis; for an organization with many products and services, it is more efficient to identify the products and services as a single comprehensive list and then audit this list customer by customer to make sure all relevant products and services are included.

The next step is not typically seen in the SIPOC model, but it is a critical part of any input/output analysis. It starts with the identification of the customers' desired outcomes—that is, the results they want as a consequence of receiving the organization's products and services. A customer who purchases a car may want years of reliable transportation, a high resale value, and styling that endures changes in vogue. A customer who purchases support services may want low-cost operations, seamless interfaces with its end-users, and a positive impact on its local community. While the organization may not have full control in helping its customers achieve these desired outcomes, it should consider (i.e., measure) how its performance contributes to or influences the achievement of these outcomes. The identification of desired outcomes also includes identifying the desired outcomes of the organization, such as financial performance (e.g., target return on investment and market share), employee retention and growth, repeat customers, and social responsibility. Measuring and comparing customer's desired outcomes to the organization's desired outcomes often highlights key management challenges, such as balancing the customer's desire for low prices with the organization's financial return targets. Measuring outcomes helps the organization understand customer needs beyond simply ensuring that outputs meet explicit specifications.

At the heart of the SIPOC model is the identification of processes, particularly the processes that produce the products and services. A separate list of support processes, those that provide internal services necessary to the functioning of the organization but are not directly involved in producing products or services for external consumption, should also be identified. Processes lend themselves to further analysis through common industrial engineering tools such as process flowcharts and value-stream maps. Process flowcharts are useful for identifying key measurement points in the flow of information and materials and thus the source of many operational performance indicators. Strategic performance measurement may include a few key process indicators, particularly those that predict the successful delivery of products and services. Once processes are identified, the inputs required for those processes are identified. As with outputs, it may be more efficient to identify inputs as a single list and then compare them to the processes to make sure all key inputs have been identified. The five generic categories of inputs that may be used to organize the list are labor, materials, capital, energy, and information. In order to be useful for identifying performance indicators, the inputs must be more specific than the five categories. For example, labor might include direct hourly labor, engineering labor, contracted labor, management, and indirect labor. These can be classified further if there is a need to measure and manage labor at a finer level, although this seems more operational than strategic. Examples of relevant labor indicators include burdened cost, hours, percent of total cost, and absenteeism. The last component of the SIPOC model is the identification of suppliers. While this component has always been important, the advent of overt improvement approaches such as supply-chain management and the increased reliance on outsourcing have made the selection and management of suppliers a key success factor for many organizations. Suppliers can also be viewed as a set of upstream processes that can be flowcharted and measured like the organization's own processes. The design team may wish to work with key suppliers to identify indicators of supplier performance that predict the success of (i.e., assure) the products and services being provided as inputs in meeting the needs of the organization's processes and subsequent products and services.

Informed by the insight of working through an input/output analysis, and regardless of whether a design team is used or not, the process of designing, developing, and implementing a measurement system based on the body of knowledge described thus far is conceptually simple and practically quite complex. An outline of the sequential steps in this process is provided as a guide in the following section.

A Macro Strategic Measurement Methodology

There are essentially seven steps in the process of building and using a strategic measurement system. Each of these seven macro steps may be decomposed into dozens of smaller activities depending on the nature and characteristics of the organization.

In practice, the steps and substeps are often taken out of sequence and may be recursive.

1. Bound the target system for which performance measures will be developed. This seemingly obvious step is included as a declaration of the importance of operationally and transparently defining the system of interest. Is the target system a single division or the entire firm? Are customers and suppliers included in the organizational system or not? Are upline policy makers who influence the environment inside or outside the system? Any particular answer may be the "right" one; the important point is shared clarity and agreement. Frequently people who want better measurement systems define a too small or limited target system, in the false belief that it is inappropriate to measure things that may be out of the target system's control. The false belief is often present at the functional and product level, and at the organizational level as supply chains become more complex. Indicators that reflect performance only partially controllable or influenced by the organization are often those most important to customers and end-users. When the organization has only partial control of a performance indicator of importance to customers, the organization needs to understand its contribution to that performance and how it interacts with factors beyond its control. This aversion to measure what is outside one's control is driven by an inability to separate measurement from evaluation. To separate the two, first, measure what is important; second, evaluate performance and the degree of influence or control users have over changing the measured result.

2. Understand organizational context and strategy. This step involves documenting, verifying, or refining the target system's mission, vision, values, current state, challenges, long- and short-term aims—all of the activities associated with strategic planning and business modeling. Recall how

to do Measurement in the Context of Planning and also the input-output process presented previously.

3. Identify the audience(s) and purpose(s) for measuring. A helpful maxim to guide development of strategic planning and strategic measurement PDSA systems is Audience + Purpose = Design. Who are the intended audiences and users of the measurement system, and what are their needs and preferences? What are the purpose(s) of the measurement system being developed? Effective measurement system designs are derived from those answers. There are many ways to discover and articulate who (which individuals and groups) will be using the measurement system, why they want to use it, and how they want to use it. Conceptually, the fundamental engineering design process is applicable here, as are the principles of quality function deployment for converting user needs and wishes into measurement system specifications and characteristics.

4. Select key performance areas (KPAs). This step involves structured, participative, generative dialogue among a group of people who collectively possess at least a minimally spanning set of knowledge about the entire target system. The output of the step is a list of perhaps seven plus or minus two answers to the following question: "In what categories of results must the target system perform well, in order to be successful in achieving its aims?"

5. For each KPA, select key performance indicators (KPIs). This step answers the question for each KPA, "What specific quantitative or qualitative indicators should be tracked over time to inform users how well the target system is performing on this KPA?" Typically a candidate set of indicators is identified for each KPA. Then a group works to clarify the operational definition and purpose of each candidate KPI; evaluate proposed KPIs for final wording, importance, data

availability, data quality, and overall feasibility; consider which KPIs will give a complete picture while still being a manageable number to track (the final "family of measures" will include at least one KPI for each KPA); select final KPIs that will be tracked; and identify the KPI "owner," sources of data, methods, and frequency of reporting, and reporting format for selected KPIs. An inventory of existing performance indicators should be completed in this step.

A note on steps 4 and 5: The order of these steps as described implies a top-down approach. However, reversing the order into a bottom-up approach can also be successful. A bottom-up approach would identify candidate indicators, perhaps using a group technique such as brainstorming or the nominal group technique (Delbecq et al., 1975). Once there is a relatively comprehensive list of candidate indicators, the list can be consolidated using a technique such as affinity diagrams (Kubiak and Benbow, 2009) or prioritized with the nominal group technique or analytical hierarchy process. The aim here is to shorten the candidate list to a more manageable size by clustering the indicators into categories that form the foundation for the dimensions of the organization's scoreboard (i.e., KPAs) or a prioritized list from which the "vital few" indicators can be extracted and then categorized by one or more of the performance dimensions frameworks to identify gaps. In either case (top-down or bottom-up), the next step is to try the indicators out with users and obtain fitness-for-use feedback.

6. Track the KPIs on an ongoing basis. Include level, trend, and comparison data, along with time-phased targets to evaluate performance and stimulate improvement. Compare and contrast seemingly related KPIs over time to derive a more integrated picture of system performance. An important part of piloting and later institutionalizing the vital few indicators is to develop appropriate portrayal formats for each indicator.

What is appropriate depends on the users' preferences, the indicator's purpose, and how results on the indicator will be evaluated. User preferences may include charts versus tables, use of color (some users are partially or fully color blind), and the ability to drill down and easily obtain additional detail. An indicator intended for control purposes must be easily transmissible in a report format and should not be dependent on color (the chart maker often loses control of the chart once it is submitted, and color charts are often reproduced on black-and-white copiers), it also should not be dependent on verbal explanation. Such an indicator should also support the application of statistical thinking so that common causes of variation are not treated as assignable causes, with the accompanying request for action. An indicator intended for feedback and improvement of the entire organization or a large group will need to be easily understood by a diverse audience, large enough to be seen from a distance, and easily dispersed widely and quickly. Design teams should support themselves with materials such as Wheeler's *Understanding Variation* (1993) and Edward Tufte's booklet, *Visual and Statistical Thinking: Displays of Evidence for Decision Making* (1997a), a quick and entertaining read on the implications of proper portrayal.

7. Conduct regular periodic face-to-face review sessions. This is a powerful approach to obtaining feedback from users on the indicators and to evaluating organizational performance based on the indicators. Review sessions are typically conducted with all the leaders of the target system participating as a group. Notionally, the review sessions address four fundamental questions: (1) Is the organization producing the results called for in the strategy? (2) If yes, what is next; if no, why not? (3) Are people completing the initiatives agreed to when deploying the strategy? (4) If yes, what is next; if no, why not? The review session is where critical thinking and group learning can occur regarding

the organizational hypothesis tests inherent in strategy. If predicted desired results are actually achieved, is it because leaders chose a sound strategy and executed it well? To what degree was luck or chance involved? If predicted results were not achieved, is it because the strategy was sound yet poorly implemented? Or was the strategy well implemented but results delayed by an unforeseen lag factor? Or, in spite of best intentions, did leaders select the "wrong" strategy? Group discussion of these strategy and measurement questions will also cause further suggestions to be made to enhance the set of indicators and how they are portrayed. See Farris et al. (2011) for more on review sessions.

PERFORMANCE MEASUREMENT PITFALLS

Performance measurement may seem rational and logical, yet implementation of many performance measurement systems fails. Here are some of the pitfalls that can contribute to failure. The reader should note that many of these pitfalls are related to the motivational aspects of measuring and evaluating performance.

- A standard set of measurements created by experts will not help. A method is needed by which measurement teams can create and continually improve performance measurement systems suited to their own needs and circumstances.

- Participation in the process of designing and implementing a performance measurement system facilitates its implementation and enhances its acceptance.

- To be "built to last," the measurement system must support decision-making and problem-solving.

- A documented and shared definition of the target system for the performance measurement effort is essential for success, as are well-crafted operational definitions for each measure of performance.

- Visibility and line-of-sight must be created for measurement systems to ensure effective utilization.

- Measurement is often resisted. Some reasons for this resistance include:

  - Data are collected but not used. It is important to be mindful that the purpose of measurement is not to generate data needlessly, but to generate data that can actually inform future decision-making.

  - Fear of the consequences of unfavorable results.

  - Fear of the consequences of favorable results, such as justifying a reduction in resources.

  - Leaders ask, "What will we do if our results are bad?" The answer is simple: You use this information as an opportunity to improve.

  - Perception that measurement is difficult.

  - If measurement activities are not integrated into work systems, they feel burdensome and like a distraction from the demands of daily business. Furthermore, measurement efforts that are not consolidated, or at least coordinated, across the organization often add unnecessary layers of complexity.

  - Measurement system design efforts are neglected.

  - In our experience, measurement is often addressed as an afterthought rather than carefully incorporated into organizational planning. Any initiative undertaken without a thoughtful planning process ultimately faces implementation challenges: measurement is no different.

  - Staff has little visibility for how measures are used.

  - Staff may not be supportive of measurement because they do not feel a connection to it or see how it can benefit them.

Integrity Audits

Performance measurements should be scrutinized, just like other functions and processes. Financial indicators and the financial control and accounting system they are typically part of receive an annual audit by an external (third-party) firm. Nonfinancial strategic performance indicators do not consistently receive the same scrutiny. So how do managers know that these nonfinancial indicators are providing them with valid, accurate, and reliable information? Valid information here refers to face or content validity: Does the indicator measure what it purports to measure? Reliable information means consistency in producing the same measurement output (i.e., indicator value) when identical performance conditions are repeatedly measured. Accuracy refers to how close the measurement output values are to the true performance values. By assuming that the indicators are providing valid, accurate, and reliable information, what assurance do managers have that their measurement systems are clearly understood, useful, and add value to the organization? A certain amount of financial measurement is a necessary part of doing business, for quarterly and annual U.S. Securities and Exchange Commission filings, reports to shareholders, or as mandated by legislation in order to continue receiving government funding. The nonfinancial components of the measurement system are not typically mandated by legislation with the exception of compliance statistics like those reported to worker safety and environmental protection agencies. Organizations compelled to participate in supplier certification programs or achieve quality or environmental management systems certification may feel coerced to develop a rudimentary nonfinancial measurement system. However, they should realize that the return from developing a strategic performance measurement system is not compliance but is the provision of useful information that adds value to the organization through better decision-making and support for implementation. After investing the time and resources to develop a strategic performance measurement system, organizations should periodically audit that system for validity, reliability, and accuracy and assess the system for continued relevance and value added.

It is beyond the scope of this chapter to describe the audit and assessment process in detail. The interested reader should refer to Coleman and Clark (2001). The "approach" includes deciding on the extent of the audit and assessment, balancing previous efforts with current needs, and choosing among the variety of techniques available.

Organizations concerned with the resource requirements to develop, operate, and maintain a measurement system may balk at the additional tasking of conducting a comprehensive audit and assessment. Such organizations should, at a minimum, subject their measurement system to a critical review, perhaps using a technique as simple as "start, stop, or continue." During or immediately following a periodic review of performance (where the current levels of performance on each key indicator are reviewed and evaluated), the manager or management team using the measurement system should ask the following three questions:

What should we start measuring that we are not measuring now? What information needs are currently unmet?

Which indicators that we are currently measuring should we stop measuring? Which are no longer providing value, are no longer relevant, or never met our expectations for providing useful information?

Which indicators should we continue to measure, track, and evaluate? If we were designing our measurement system from scratch, which of our current indicators would appear again?

Another less resource-intensive approach is to address the auditing and assessing of the measurement system as part of a periodic organizational assessment.

ORGANIZATIONAL ASSESSMENTS: STRATEGIC SNAPSHOTS OF PERFORMANCE

Organizational assessments are a periodic snapshot form of strategic performance measurement. They are periodic in that they do not measure performance frequently: once a year to once every 5 or 10 years is common. They are snapshots because they reflect the organization's performance at a particular time and may not be fully evaluated until several weeks or months later. They are relatively comprehensive in scope, often measuring and evaluating all or most of the enterprise's activities and results, including the organization's measurement and evaluation system. Preparing for an organizational assessment may require a review of the organization's measurement system, and the assessment process will provide both direct and indirect feedback on the usefulness and value of the measurement system. Organizational assessments are used for conformity, to ensure the organization meets some standard (e.g., accreditation or certification), or for improvement and recognition where the organization is compared with a standard and provided feedback for improvement. Those exhibiting the highest levels of performance against the standard are recognized with an organizational award (e.g., Baldrige Award, State or Corporate Awards for Excellence, EFQM Excellence Award).

Organizational assessment typically begins with a self-study comparing the organization and its goals against an established standard (i.e., criteria or guidelines). The completed self-study is then submitted to a third party (i.e., the accreditation, registration, or award body) for review and evaluation. This third-party review begins with an evaluation of the self-study and is often, but not always, followed by a visit to the organization. The purpose of the visit is to validate and clarify what was reported in the self-study. The third party then renders a judgment and provides feedback to the organization. Depending on the specific application, the third-party judgment may result in substantial consequences for the organization (e.g., winning an award, receiving accreditation, or failure to do so). Ideally, the feedback from the third party is fed into the organization's improvement cycle, implemented, measured, and reflected in future plans and results.

Organizations that operate an ongoing improvement cycle and feed the results of the assessment into that cycle are likely to receive the greatest return on the investment from the resources required to complete the self-study and assessment. Particularly in situations where the organizational assessments occur several years apart, having an ongoing improvement process maintains the momentum and focus on what is important and should make preparing for future assessments easier. The improvement process translates assessment findings into plans, actions, and targets, applies resources, and then follows up with regular review of results and then new or updated plans, actions, and targets. While the overall improvement process should be management-led, industrial engineers are often tasked as analysts and project managers to convert assessment findings into plans, actions, and results.

Organizations wishing to gain much of the benefit of a comprehensive assessment but concerned about the resource requirements should simply complete a five-page organizational profile, the preface of a Baldrige Award application (self-study) (Baldrige Performance Excellence Program, 2011, pp. 4–6). The organizational profile asks the organization to document its organizational environment including product offerings, vision and mission, workforce, facilities, technologies, equipment, and regulatory requirements; its organizational relationships including organization structure, customers and stakeholders, suppliers and partners; its competitive environment including competitive position(s), competitiveness changes, and comparative data for evaluating performance; its strategic context in terms of key business, operational, social responsibility, and human-resource challenges and advantages; and a description of its performance improvement system. For many organizations, particularly, smaller organizations and departments or functions within larger organizations, developing and collectively reviewing the organizational profile may provide more than 50 percent of the value of a complete organizational assessment. Too few management teams have developed consensus answers to the questions posed by the organizational profile. Developing the organizational profile as a team and keeping it current provides a key tool for providing organizational direction and furnishes an important input into the development and maintenance of the performance measurement system. Even organizations not interested in the Baldrige or other business excellence awards can use the profile as a resource for the development of management systems or the preparation of a self-study.

Organizational assessments, like other forms of performance measurement, should be subject to periodic audit and assessment. The reliability and validity of the results of organizational assessments are not as well investigated as we might like. Few, if any, of the organizations that offer or manage these assessments provide statistics showing they periodically evaluate the efficacy of their assessment processes. Researchers (Coleman et al., 2001, 2002; Coleman and Koelling, 1998; Keinath and Gorski, 1999) have estimated some of the properties associated with the scores and feedback received from organizational assessments. Their findings suggest that training the assessors (a.k.a. evaluators or examiners) reduces scoring leniency; however, their findings are less conclusive regarding the effect of training on inter-rater reliability and accuracy. Those interested in interpreting the variability observed among results from organizational assessments should consult the above-cited sources.

REFERENCES

Akao, Y. (Ed.). 1990, *QFD: Quality Function Deployment*. Cambridge, MA: Productivity Press.

Baldrige Performance Excellence Program. 2011, *2011–2012 Baldrige Criteria for Performance Excellence*. Gaithersburg, MD: National Institute of Standards and Technology.

Barrett, R., March 4, 1999a, *Liberating the Corporate Soul*. Arlington, VA: Presentation to The Performance Center.

Barrett, R. 1999b, *Liberating the Corporate Soul*. Alexandria, VA: Fulfilling Books.

Becker, B., Huselid, M. A. and Ulrich, D. 2001, *The HR Scorecard: Linking People, Strategy and Performance*. Boston, MA: Harvard Business School Press.

Bititci, U., Garengo, P., Dörfler, V., and Nudurupati, S. 2012, Performance Measurement: Challenges for Tomorrow, *International Journal of Management Reviews*, Vol. 14, No. 3. pp. 305–327.

Bourne, M., Neely, A., Platts, K. and Mills, J. 2002, The Success and Failure of Performance Measurement Initiatives: Perceptions of Participating Managers, *International Journal of Operation and Production Management*, Vol. 22, pp. 1288–1310.

Brown, M. G. 1996, *Keeping Score: Using the Right Metrics to Drive World-Class Performance*. New York, NY: AMACOM Books.

Brown, M. G. 2000, *Winning Score: How to Design and Implement Organizational Scorecards*. Portland, OR: Productivity Press.

Busi, M. and Bititci, U. S. 2006, Collaborative Performance Management: Present Gaps and Future Research, *International Journal of Productivity and Performance Management*, Vol. 55, pp. 7–25.

Coleman, G. D. and Clark, L. A. 2001, A Framework for Auditing and Assessing Non-Financial Performance Measurement Systems, in *Proceedings of the Industrial Engineering Research Conference*. [CD-ROM].

Coleman, G. D. and Clark, L. A. 2014, Strategic Performance Measurement, in Badiru, A. B. (Ed.), *Handbook of Industrial and Systems Engineering*. Boca Raton, FL: Taylor and Francis Group.

Coleman, G. D., Costa, J. and Stetar, W. 2004, The Measures of Performance: Managing Human Capital Is a Natural Fit for Industrial Engineers, *IE: Industrial Engineering*, Vol. 36, pp. 40–44.

Coleman, G. D. and Koelling, C. P. 1998, Estimating the Consistency of Third-Party Evaluator Scoring of Organizational Self-Assessments, *Quality Management Journal*, Vol. 5, pp. 31–53.

Coleman, G. D., Koelling, C. P. and Geller, E. S. 2001, Training and Scoring Accuracy of Organisational Self-Assessments, *International Journal of Quality Reliability Management*, Vol. 18, pp. 512–527.

Coleman, G. D., Van Aken, E. M. and Shen, J. 2002, Estimating Interrater Reliability of Examiner Scoring for a State Quality Award, *Quality Management Journal*, Vol. 9, pp. 39–58.

Cooper, W. W., Seiford L. M., and Zhu, J. (Eds.). 2004, *Handbook on Data Envelopment Analysis*. New York, NY: Springer.

Deal, T. E. and Kennedy, A. 1982, *Corporate Cultures*. Reading, MA: Addison-Wesley.

Delbecq, A. L., Van De Ven, A. H. and Gustafson, D. H. 1975, *Group Techniques for Program Planning: A Guide to Nominal Group and Delphi Processes*. Middleton, WI: Green Briar.

Deming, W. E. 1960, *Sample Design in Business Research*. New York, NY: Wiley.

Deming, W. E. 1993, *The New Economics for Industry, Government, and Education*. Cambridge, MA: MIT Center for Advanced Engineering Study.

Drucker, P. F. 1954, *The Practice of Management*. New York, NY: Harper and Row.

Drucker, P. F. 1998, The Next Information Revolution, *Forbes ASAP*, pp. 47–58.

Farris, J. A., Van Aken, E. M., Letens, G. Chearksul, P. and Coleman, G. 2011, Improving the Performance Review Process: A Structured Approach and Case Application, *International Journal of Operations and Production Management*, Vol. 31, pp. 376–404.

Garvin, D. A. 1984, What Does "Product Quality" Really Mean?, *Sloan Management Review*, Vol. 26, pp. 25–43.

GOAL/QPC. 1985, *The Memory Jogger*, Salem, NH: GOAL/QPC.

Hamel, G. 2002, *Leading the Revolution*. Cambridge, MA: Harvard Business School Press. (Revised).

Hamel, G. and Prahalad, C. K. 1996, *Competing for the Future*. Cambridge, MA: Harvard Business School Press.

Ishikawa, K. 1985, *What Is Total Quality Control? The Japanese Way* (trans. Lu, D. J.). Englewood Cliffs, NJ: Prentice Hall.

Kaplan, R. S. and Norton, D. P. 1992, The Balanced Scorecard: Measures That Drive Performance, *Harvard Business Review*, Vol. 70, pp. 71–79.

Kaplan, R. S. and Norton, D. P. 1996, Using the Balanced Scorecard as a Strategic Management System, *Harvard Business Review*, Vol. 74, pp. 75–85.

Keinath, B. J. and Gorski, B. A. 1999, An Empirical Study of the Minnesota Quality Award Evaluation Process, *Quality Management Journal*, Vol. 6, pp. 29–38.

Kennerley, M. and Neely, A. 2002, A Framework of the Factors Affecting the Evolution of Performance Measurement Systems, *International Journal of Operations and Production Management*, Vol. 22, pp. 1222–1245.

Kilmann, R. 1989, *Managing Beyond the Quick Fix*. San Francisco, CA: Jossey-Bass.

Kilmann, R. and Saxton, M. J. 1983, *Kilmann-Saxton Culture Gap Survey*. Tuxedo, NY: XICOM.

Kizilos, T. 1984, Kratylus Automates His Urnworks, *Harvard Business Review*, Vol. 62, pp. 136–144.

Kubiak, T. M. and Benbow, D. W. 2009, *The Certified Six Sigma Black Belt Handbook*, 2nd ed. Milwaukee, WI: Quality Press.

Lawton, R. 2002, Balance Your Balanced Scorecard, *Quality Progress*, Vol. 35, pp. 66–71.

Leedy, P. D. and Ormrod, J. E. 2001, *Practical Research: Planning and Design*. Upper Saddle River, NJ: Merrill Prentice Hall.

Mallak, L. A., Bringelson, L. S. and Lyth, D. M. 1997, A Cultural Study of ISO 9000 Certification, *International Journal of Quality and Reliability Management*, Vol. 14, pp. 328–348.

Mallak, L. A. and Kurstedt, H. A. 1996, Using Culture Gap Analysis to Manage Organizational Change, *Engineering Management Journal*, Vol. 8, pp. 35–41.

Medina-Borja, A. M., Pasupathy, K. S. and Triantis, K. 2006, Large-Scale Data Envelopment Analysis (DEA) Implementation: A Strategic Performance Management Approach, *Journal of the Operational Research Society*, Vol. 58, pp. 1084–1098.

Miller, D. 1984, Profitability = Productivity + Price Recovery, *Harvard Business Review*, Vol. 62, pp. 145–153.

Muckler, F. A. and Seven, S. A. 1992, Selecting Performance Measures: "Objective" Versus "Subjective" Measurement, *Human Factors*, Vol. 34, pp. 441–455.

Neely, A. 1999, The Performance Measurement Revolution: Why Now and What Next?, *International Journal of Operations and Production Management*, Vol. 19, pp. 205–228.

Neely, A., Bourne, M. and Adams, C. 1999, Better Budgeting or Beyond Budgeting? *Measuring Business Excellence*, Vol. 7, pp. 22–28.

Nørreklit, H. 2003, The Balanced Scorecard: What Is the Score? A Rhetorical Analysis of the Balanced Scorecard, *Accounting, Organizations and Society*, Vol. 28, pp. 591–619.

Peters, T. J. and Waterman, R. H. Jr 1982, *In Search of Excellence: Lessons from America's Best-Run Companies*. New York, NY: Warner Books.

Pfeffer, J. and Sutton, R. I. 2000, *The Knowing-Doing Gap: How Smart Companies Turn Knowledge into Action*. Boston, MA: Harvard Business School Press.

Pineda, A. J. 1996, Productivity Measurement and Analysis (Module 4), in Prokopenko, J. and North, K. (Eds.), *Productivity and Quality Management: A Modular Programme*. Geneva, Switzerland: International Labor Organization.

Rucci, A. J., Kirn, S. P. and Quinn, R. T. 1998, The Employee-Customer-Profit Chain at Sears, *Harvard Business Review*, Vol. 76, pp. 82–97.

Sink, D. S. 1985, *Productivity Management: Planning, Measurement and Evaluation, Control and Improvement*. New York, NY: Wiley.

Sink, D. S. and Tuttle, T. C. 1989, *Planning and Measurement in Your Organization of the Future*. Norcross, GA: IIE Press.

Sumanth, D. J. 1998, *Total Productivity Management: A Systematic and Quantitative Approach to Compete in Quality, Price, and Time*. Boca Raton, FL: St. Lucie Press.

Thompson, A. A., Jr. and Strickland A. J, III. 2003, *Strategic Management: Concepts and Cases*, 13th ed. Boston, MA: Irwin/McGraw-Hill.

Thor, C. G. 1998, *Designing Feedback*. Menlo Park, CA: Crisp Publications.

Tufte, E. R. 1997a, *Visual and Statistical Thinking: Displays of Evidence for Decision Making*. Cheshire, CT: Graphics Press.

Tufte, E. R. 1997b, *Visual Explanations: Images and Quantities, Evidence and Narrative*. Cheshire, CT: Graphics Press.

Wheeler, D. J. 1993, *Understanding Variation: The Key to Managing Chaos*. Knoxville, TN: SPC Press.

CHAPTER 8

Industrial Engineering and Lean Manufacturing

T HE GOAL OF INDUSTRY is to operate clean and lean. One of the major achievements of industrial engineering is helping industry achieve lean operations. Badar (2014) presents the processes for achieving lean manufacturing cells. Lean principles focus on elimination of waste by applying tools such as just-in-time (JIT), level production, standardized work, quality at the source, and continuous improvement. This chapter discusses the application of lean concepts to design and organize a manufacturing cell. A brief description of the system engineering approach as well as a summary of common sources of waste are also presented. A methodology to organize a cell using a 6S (sort, straighten, shine, standardize, safety, and sustain) program along with a point system that can be designed to account for the 6S is explained in detail. An audit of the cell will yield a score. Making necessary changes taking waste reduction into consideration will improve the score. Periodic audits should be conducted to assure continuous improvement.

Lean manufacturing is a set of principles, concepts, and techniques derived from Toyota's JIT production system (Monden, 1998; Parks, 2003). *Just-in-time* means achieving the level of production that precisely and flexibly matches customer demand and consists of processes that employ minimal (ideally zero) inventory through a strategy where each operation supplies parts or products to successor operations at the precise time they are demanded. It requires a continuous flow process structure that uses multifunction employees performing only value-added operations. It employs a pull system (kanban system) strategy to meet demand and limit in-process inventories. This system is known as "lean" because of its ability to do so much more with fewer resources (space, machines, labor, and materials) than traditional approaches. It uses standardized work practices based on minimal workforce and effort, highest quality, and highest safety in performing each job. It emphasizes building quality into the product rather than inspecting quality through systems that identify and resolve quality problems at their source. Thus, it provides customers with more value and company with less waste (Durham, 2003). Waste can be due to any of the following: overproduction, motion, transportation, waiting, setup time, processing time, inventory, defective products, and underutilized workers (Askin and Goldberg, 2002; Minty, 1998).

The objectives of lean manufacturing are as follows (Sobek and Jimmerson, 2003): to use as few resources as possible, to produce the desired amount of product at the highest possible level of quality, and in as short period of time as possible (reducing wait time that occurs as materials wait in queue or in inventory, and decreasing setup time). The key to doing this is to produce in small batch sizes. As inventory levels fall, the cost of defects soars because the system has little slack to absorb them. Thus, great attention is paid to fixing problems if defects occur. Also, work processes must be finely tuned and standardized to achieve predictable processing times and quality. The result is, to the extent possible in a mass-production environment, a system that focuses on individual products made for individual customers.

In recent years, lean philosophies and practices have been implemented in several manufacturing facilities in the United States and around the world with such success that it is rapidly becoming the dominant manufacturing paradigm. Lean manufacturing integrates simple low-tech tools with advanced production/information technology and unique social/management practices (Sobek and Jimmerson, 2003). This organizational aspect makes it different from total quality management (TQM), which is rooted in meeting customer satisfaction or trying to meet the Six-Sigma (Elliott, 2003) requirements (3.4 parts per million [ppm] defective or 99.999% perfection). But TQM efforts often do not address organizational systems well and are not responsive to the needs of the employees and management. Integrated with management tools, lean manufacturing can help ensure the achievement of a company's strategic objectives.

WASTE ELIMINATION

Lean manufacturing principles emphasize the elimination of waste from a production system in order to make it more efficient (Pondhe et al., 2006). In reality, however, such waste is generally hidden. Before removing waste, it is important to identify its sources (Askin and Goldberg, 2002; Minty, 1998), which are as follows:

Overproduction: Production of any product involves costs associated with direct material, direct labor, and manufacturing overhead, which can include factory overhead, shop expense, burden, and indirect costs. This means that the quantity produced per period of time from a manufacturing cell should be set to match the demand so that all the items made can be sold. Production should never be set to keep the resources busy. Any amount over the demand is waste as it costs money as well as wearing of machines.

Motion: Motion associated with either human motion or material handling consumes time and energy, and any motion that does not add value is a waste. Therefore, workplace and corresponding processes should be designed to eliminate non-value-added motions and to include ergonomic and safety considerations.

Transportation: This includes the movement between work-cell and storage area. Excessive movements should be minimized. Toolkits can be placed close to the point of use (POU). Materials can be stored and oriented in such a way so that they can be fed to the cell easily.

Waiting: If a material or work in process (WIP) is waiting in queue for a proper machine or worker to be available, this is a kind of waste and causes longer throughput time. Production in small batches with coordinated order processing reduces excessive WIP and cycle time.

Setup time: Every time a tool setup is changed, it requires motion, time, and energy. A workplace should be designed to minimize the number of setups.

Processing time: A production system may consist of value-added and non-value-added activities. Processing time can be reduced by avoiding nonvalue operations.

Inventory: Inventory of finished goods involves costs of space, obsolescence, damage, opportunity cost, and handling. Therefore, excessive inventory should be eliminated.

Defective products: Defective products cause two problems: cost of material and resource, and poor customer satisfaction. Therefore, quality of the system, process, and products should be monitored continually to decrease defective products.

Underutilized workforce: If people working in a manufacturing cell or a production system are not utilized completely, meaning there is not enough work for all of them, this is also a waste.

SYSTEMS VIEW OF LEAN

A system engineering approach can help industry design a manufacturing facility that will be not only functionally highly productive but also good from the viewpoint of flow of materials or processes. The idea is to consider the manufacturing facility as a system (a set of interacting parts). In the case of the manufacturing facility, different cells are the interacting parts of the system. For any system, adequate stakeholders should be identified and listed (Sawle et al., 2005)—"stakeholders" being persons or organizations that directly have "something at stake" in the creation or operation of the subject system. Examples may include company owners or management, shareholders, employees, the local community, customers, retailers, etc. After the identification and listing of stakeholders, features of the subject system should be examined and enumerated (Sawle et al., 2005). A feature is a behavior of the system that has value to the stakeholders, each feature consisting of one or more feature attributes. For example, for a manufacturing system, production may be a feature and its attributes may include production rate and cost. This should be supplemented with a domain diagram (Sawle et al., 2005) of the system providing a high-level view of the environment in which the system exists and interacts with surrounding subsystems. Functional requirements of the system can be derived from these interactions (Sawle et al., 2005). Further, the feature- and role-attribute mapping will provide an idea of the optimum combination of design attribute values and stakeholder feature attributes (Sawle et al., 2005). This will help decide on feature attributes or parameters to be concentrated on for improvement and investment.

LEAN MANUFACTURING CELL

In the preceding sections, lean concepts and system engineering techniques have been described. A manufacturing system and its corresponding cells should be designed considering all these principles in order to efficiently create value for its multiple stakeholders. This requires deep commitment from the top management.

A manufacturing cell consists of a particular group of operations. For instance, a cell may contain hole-punching operations. It is important to note that a cell may be related to other cells in terms of the flow of materials and processes. Thus, each cell and its corresponding operations or processes and subprocesses need to be designed efficiently.

The first step toward good design is to describe and name all the processes and subprocesses. The second is to measure performance data such as time taken and resources (material, machine, labor, space, etc.) being utilized over a period of time. The third is to analyze the existing process. Why is each process and subprocess necessary? Do all employees follow the same sequence of events prescribed for the process? How much is the process variation? How much is the utilization of each resource? Cause–effect analysis should be performed. Feedback from management, employees, and clients or customers should be obtained. A scoring system that is weighted to assess the importance of each task should be utilized. The fourth step is to propose necessary changes to improve the process. The fifth is to implement the suggested changes for improvements. The sixth is to evaluate (control, standardize, and verify) the new process. Last is continuous improvement. Otherwise, everything will go back to the old state. The steps for designing a process are as follows:

Step 1: Name the process and subprocesses

Step 2: Measure performance data

Step 3: Analyze the process

Step 4: Suggest changes to improve the process

Step 5: Implement changes

Step 6: Standardize and verify the process

END: Continuous improvement

A lean cell needs to be designed to meet customer demands. Depending on the demand, the workforce can be increased or decreased. But operations or tasks within a work-cell must be set in such a way that it should be able to adjust to the workforce level.

ORGANIZING A CELL

Lean concepts can also be implemented to organize a workplace using 5S philosophy (Brian, 2003; Parks, 2003; Standard and Davis, 1999):

Sort and clear out: Eliminate what is not needed, that is, remove non-value-added processes or actions that increase the cost but add no value to the product.

Straighten and configure: A place for everything (materials and tools) and everything in its place, so that they are easily accessible.

Scrub, shine, and clean up: Clean the workplace and look for ways to keep it clean. Clean work environments are more productive and provide the workers with less stressful environments.

Safety and self-discipline: Always think safety first. Safety is very important in order to provide workers with satisfactory working conditions.

Standardization: Standardized work methods eliminate variation and non-value-added time in a process.

Some companies add one more S to make a 6S system (Badar and Johnston, 2004): sustain or stick to the rules, which include maintaining and monitoring the above practices.

Badar and Johnston (2004) have investigated and carried out the implementation of lean concepts or 6S to organize a cell of a local manufacturing company *X*. It was started with three basic steps (Durham, 2003): walk around, get creative, and look beyond the four walls. The cell consisted of one H5 Cincinnati (a five-axis CNC controlled mill) and two Omni mills. The procedure is described in the following section.

PROCEDURE

Organization of the workplace is the beginning of lean application. This can be done using a 6S program: sort, straighten, shine, standardize, safety, and sustain. In Badar and Johnston (2004), a point system was designed to account for each category of 6S: sort, 25 points; straighten, 25 points; shine, 25 points; standardize, 25 points; safety, 30 points; and sustain, 25 points. In addition, 30 points were assigned for the environmental and ISO 14001 consideration, yielding a total of 185 maximum points. Each of the seven categories was further broken down into five subcategories. For each subcategory, three to six possible points were assigned. Thus, there were 35 subcategories and 185 possible points in total. To ensure the privacy and confidentiality of the company, no further details can be specified.

With this point system in place, the plan was to conduct an audit of the existing workplace; it would be considered acceptable if it scored 148 out of 185 possible points. A score above 164 was needed to classify the cell area in the Green category as per the company standard. The goal was to keep auditing and organizing the cell using lean concepts so that finally it could be classified into the Green category.

An initial audit of the cell was conducted and the cell scored 133 points, below the acceptable level. The categories and subcategories receiving low points were identified. The issue was discussed with the employees to get their input. As outsiders, the authors (Badar and Johnston, 2004) were able to see beyond their wall and give unbiased feedback. This input helped determine what steps were needed, and in what order, for the implementation of the 6S program.

The following steps were taken to organize the cell:

1. The aisles were painted so that no material could be stacked in the aisles. This would also help control the stock on the floor of the cell area better.

2. An attempt was made to investigate what material and tools the machine operators were using and what they were not. Anything that was not being used was removed from the cell.

3. After thinning out the unused items, a place for each item (material and tool) was made and identified, so that anyone working in the cell could readily find what he or she needed. For example, squares were painted on the floor for the inspection cart and the operator's tool chest. This would also help prevent overstocking of materials, as it would be seen by the workers.

4. The floor, machine, and tools were cleaned up.

5. Labels and signs were posted to ensure safety and keep the cell organized.

With the above changes made, an audit was conducted and the cell scored 175 points. This score classified the cell into the Green category. However, periodic audits should be conducted to keep the process of continuous improvement in place.

SIMULATION MODELING

A simulation model of a manufacturing cell in Arena or other similar software can be developed to provide visual aspects to the complexities of the operations and resources associated with the cell (Thomson and Badar, 2004). The model is a simple yet effective approach to the problem-solving process. It can provide answers that could otherwise not be seen, because the simulation model can be modified and variables changed without affecting the current production. Although the accuracy of the results is as accurate as the data input, the Arena software contains an internal randomizer that can provide accurate statistical data.

Simulation can quantify the performance improvements that are expected from implementing the lean manufacturing philosophy of continuous flow, JIT inventory management, quality at the source, and level production scheduling. Results from the simulation model may be gathered quickly. Detty and Yingling (2000) have used simulation to assist in the decision to apply lean methods at an existing assembly operation. The manufacturing system in their study composed multiple identical cells that assembled a finished product from over 80 individual parts. The cells were either fully staffed or shut down as product demand varied. In their study, they built models for the existing system as well as for a new system with the implementation of lean principles. Their models demonstrated the entire manufacturing system, including the manufacturing processes and the associated warehousing, in-process inventory levels, transportation, and production-control and scheduling systems. Thus, the simulation model helped quantify the benefits of the lean manufacturing concepts.

CONCLUSION

A manufacturing cell designed by employing lean principles and system-engineering approaches efficiently creates value for its multiple stakeholders. Organization of a cell using the 6S system means that every item will be in its proper place and the workplace will be clean and shining. The 6S philosophy, combined with employee input, common sense, and looking "beyond the walls," can help reduce waste and result in an efficient workplace. Owing to global warming and other environmental problems, an environmental consideration should be made from the beginning and in every aspect of manufacturing. It is to be noted that continuous improvement is an essential concept of lean philosophy. Therefore, once a cell or a system is designed and organized, it is a must to continuously improve safety, quality, productivity, and the work environment.

REFERENCES

Askin, R. G. and Goldberg, J. B. 2002, *Design and Analysis of Lean Production Systems*. New York, NY: Wiley.

Badar, M. A. 2014, Lean Manufacturing Cell, in A. B. Badiru (Ed.), *Handbook of Industrial and Systems Engineering*. Boca Raton, FL: Taylor and Francis Group.

Badar, M. A. and Johnston, C. 2004, Lean Application in Organizing a Manufacturing Cell, in *Proceedings of the IIE Annual Conference 2004, Solutions* [CD-ROM], Houston, TX, May 15–19.

Brian, S. 2003, Case Study in Creating a Work Cell – JIT, TPM and 5S Implementation, in *IIE 2003 Solutions Conference, Track: Lean Concepts and Applications*, Portland, OR, May 17–21.

Detty, R. B. and Yingling, J. C. 2000, Quantifying Benefits of Conversion to Lean Manufacturing with Discrete Event Simulation: A Case Study, *International Journal of Production Research*, Vol. 38, No 2, pp. 429–445.

Durham, D. R. 2003, Lean and Green: Improving Productivity and Profitability, *Manufacturing Engineering*, Vol. 131, No 3, p. 16.

Elliott, G. 2003, The Race to Six Sigma, *Industrial Engineer*, Vol. 35, No. 10, pp. 30–34.

Minty, G. 1998, *Production Planning and Controlling*. Tinley Park, IL: Goodheart-Willcox.

Monden, Y. 1998, *Toyota Production System: An Integrated Approach to Just-in-Time*, 3rd ed. Norcross, GA: Institute of Industrial Engineers.

Parks, C. M. 2003, The Bare Necessities of Lean, *Industrial Engineer*, Vol. 35, No. 8, pp. 39–42.

Pondhe, R., Asare, S. A., Badar, M. A., Zhou, M., and Leach, R. 2006, Applying Lean Techniques to Improve an Emergency Department, in *Proceedings of the IIE Annual Conference 2006, Session: IERC03 Engineering Management 6* [CD-ROM], Orlando, FL, May 20–24.

Sawle, S. S., Pondhe, R., Asare, S. A., Badar, M. A., and Schindel, W. 2005, Economic Analysis of a Lawn Mower Manufacturing System Based on Systems Engineering Approach, *IIE Annual Conference 2005, Research Track: Engineering Economics*, Atlanta, GA, May 14–18.

Sobek, D. K. and Jimmerson, C. 2003, Innovating Health Care Delivery Using Toyota Production System Principles, A Proposal to the National Science Foundation Innovation and Organizational Change Program, retrieved on September 21, 2003, from http://www.coe.montana.edu/ie/faculty/sobek/IOC_Grant/proposal.htm.

Standard, C. and Davis, D. 1999, *Running Today's Factory: A Proven Strategy for Lean Manufacturing*. Cincinnati, OH: Hanser Gardner Publications.

Thomson, B. A. and Badar, M. A. 2004, Receiving Area Improvement Using Lean Concepts and Simulation, in *Proceedings of the 2004 NAIT Selected Papers* [CD-ROM], Louisville, KY, October 19–23.

CHAPTER 9

Industrial Engineering and Human Factors

INDUSTRIAL ENGINEERING IS A human-centric discipline. As such, human factors are a key part of the research and practice of industrial engineering. Resnick (2014) presents the basics of human factors in the industrial engineering context.

Human Factors is a science that investigates human behavioral, cognitive, and physical abilities and limitations in order to understand how individuals and teams will interact with products and systems. Human Factors engineering is the discipline that takes this knowledge and uses it to specify, design, and test systems to optimize safety, productivity, effectiveness, and satisfaction.

Human Factors is important to industrial and systems engineering because of the prevalence of humans within industrial systems. It is humans that, for the most part, are called on to design, manufacture, operate, monitor, maintain, and repair industrial systems. In each of these cases, Human Factors should be used to ensure that the design will meet system requirements in performance, productivity, quality, reliability, and safety. This chapter presents an overview of Human Factors, how it should be integrated into the systems engineering process, and some examples from a variety of industries.

The importance of including Human Factors in systems design cannot be overemphasized. There are countless examples that illustrate the importance of Human Factors for system performance. A 1994 survey found that 92 percent of computer-related fatalities between 1979 and 1992 could be attributed to failures in the interaction between a human and a computer. The extent of the 1979 accident at the Three Mile Island nuclear power plant was largely due to Human Factors challenges (Bailey, 1996), almost resulting in a disastrous nuclear catastrophe. The infamous butterfly ballot problem in Florida in the 2000 U.S. presidential election is a clear example of an inadequate system interface yielding remarkably poor performance (Resnick, 2001). Web sites such as www.baddesigns.com, http://goodexperience.com/tib/, and others provide extensive listings of designs from everyday life that suffer from poor consideration of Human Factors. Neophytes often refer to Human Factors as common sense. However, the prevalence of poor design suggests that Human Factors sense is not as common as one might think. The consequences of poor Human Factors design can be inadequate system performance, reduced product sales, significant product damage, and human injury.

This chapter provides an overview of Human Factors and is intended to support the effective design of systems in a variety of work domains, including manufacturing, process control, transportation, medical care, and others. The section entitled The Benefits of Human Factors presents some of the principal components of Human Factors analysis that must be addressed in any systems design and the benefits of integrated effective Human Factors. The section entitled A Human Factors Conceptual Model describes a conceptual model of human information processing and outlines how each aspect affects performance. An example is provided for each one that illustrates the design challenges for Human Factors and how they can be overcome. The section Cognitive Consequences of Design describes two important consequences of design: the ability of humans to learn from their experience and the likelihood of error during system use.

ELEMENTS OF HUMAN FACTORS

In order to facilitate the design of effective systems, Human Factors must adopt a holistic perspective on human-system interaction. Systems engineers need to understand how people think, how these thoughts lead them to act, the results of these actions, and the reliability of the results of these actions. Thus, the following four elements should be considered: cognition, behavior, performance, and reliability.

Cognition

A considerable body of Human Factors research has been dedicated to human cognition. It is critical for systems engineers to understand and predict how users will perceive the information that they receive during system use, how this information will be processed, and the nature of users' resulting behavior and decisions. Situation awareness (Endsley, 2000a) refers to the extent to which a user has perceived and integrated the important information in the world and can project that information into the future to make predictions about system performance.

Consider the case of an air-traffic controller who needs to monitor and communicate simultaneously with several aircraft to ensure they all land safely. This job requires the controller to develop a composite mental model of the location and direction of each aircraft so that when new information appears, he or she can quickly decide on an appropriate response. The design of the system interface must anticipate this model so that information can be presented in a way that allows the controller to perceive it quickly and effectively integrate it into the mental model.

Behavior

The actions taken by the human components of a system are often more difficult to predict than the mechanical or electrical components. Unlike machines, people behave based on experiences and beliefs that transcend the system, including factors such as corporate culture, personal goals, and past experience. It is critical for systems engineers to investigate the effects of these sources on behavior to ensure that the system will be successful.

For example, the Columbia Accident Investigation Board (CAIB) concluded that the accident causing the destruction of the space shuttle *Columbia* in 2003 was as much caused by the NASA organizational culture as it was by the foam that struck the orbiter. The CAIB report stated that systems were approved despite deviations in performance because of a past history of success (CAIB, 2003). At the consumer level, various factors may also be important. For instance, Internet retailers are interested in the factors that determine whether a consumer will purchase a product on the company's Web site. In addition to the design factors, such as the site's menu design and information architecture, the user's past history at other Web sites can also affect his or her behavior on this site (Nielsen, 1999).

Performance

Most systems depend not only on whether an action is completed, but also on the speed and accuracy with which the action is completed. Many factors affect user performance, such as the number of information sources that must be considered, the complexity of the response required, the user's motivation for performing well, and others (Sanders and McCormick, 1993).

Call-center operations are a clear example of the need to include Human Factors in design to achieve optimal performance (Kemp, 2001). Call-center software must complement the way that operators think about the task, or performance may be significantly delayed. The cost structure of call centers relies on most customer service calls being completed within seconds. Early versions of some customer relationship management (CRM) software required operators to drill down through 10 screens to add a customer record. This design slowed the task considerably. However, labels that lead to strong path recognition can have as great an effect as path length on performance (Katz and Byrne, 2003). Trade-offs between path recognition strength and path length must be resolved in the information architecture of the system. It is critical that systems that rely on speed and accuracy of performance thoroughly integrate Human Factors into their designs.

Reliability

Human Factors is also important in the prediction of system reliability. Human error is often cited as the cause of system failures (FAA, 1990). However, the root cause is often traceable to an incompatibility between the system interface and human information processing. An understanding of human failure modes, the root causes of human error, and the performance and contextual factors that affect error probability and severity can lead to more reliable systems design.

Much of the research in human error has been in the domain of aerospace systems and control center operations (Swain and Guttman, 1983). The open literature contains models that predict human errors in behavior, cognition, communication, perception, and other human functions. Integrating human reliability models into the systems engineering process is essential.

THE BENEFITS OF HUMAN FACTORS

There are many benefits that result from considering each of these four elements of Human Factors in systems design. The primary benefit is that the resulting system will be more effective. By accommodating the information-processing needs of the users, the system will better match the system requirements and will thus be more productive. Systems that incorporate Human Factors are also more reliable. Since human error is often the cause of system failure, reducing the likelihood of human error will increase the reliability of the system as a whole.

Consideration of Human Factors also leads to cost reductions in system design, development, and production. When Human

Factors are considered early in the design process, flaws are avoided and early versions are closer to the final system design. Rework is avoided, and extraneous features can be eliminated before resources are expended on developing them.

Human Factors also leads to reduced testing and quality assurance requirements. Problems are caught earlier, and it becomes easier to prioritize what system components to modify. Systems that exhibit good Human Factors design reduce sales time and costs because they are easier to demonstrate, train, and set up in the field.

Finally, consideration of Human Factors leads to reduced costs for service and support. When systems are easy to use, there are fewer service calls and less need for ongoing training. The occurrence of fewer errors leads to reduced maintenance costs, fewer safety violations, and less frequent need for mishap/injury investigation.

A HUMAN FACTORS CONCEPTUAL MODEL

Behavior and performance emerge from the way that humans process information. Human information processing is generally conceptualized using a series of processing stages. It is important to keep in mind that these stages are not completely separate; they can work in parallel, and they are linked bidirectionally. A detailed discussion of the neurophysiology of the brain is beyond the scope of this chapter. But there is one underlying trait that is often overlooked. The human information-processing system (the brain) is noisy, a fact that can lead to errors even when a particular fact or behavior is well known. On the positive side, this noise also enables greater creativity and problem-solving ability.

Long-Term Memory

Long-term memory refers to the composite of information that is stored in an individual's information-processing system, the brain. It is composed of a vast network of interconnected nodes, a network that is largely hierarchical but that has many cross-unit connections as well. For example, a dog is an animal but also lives (usually) in a house.

The basic unit of memory has been given many names but for the purposes of this chapter will be called the *cell assembly* (Hebb, 1955). A cell assembly is a combination of basic attributes (lines, colors, sounds, etc.) that become associated because of a history of being activated simultaneously. Cell assemblies are combined into composites called *schema*. The size and complexity of the schema depend on the experience of the individual. The schema of an elephant will be very simple for a four-year-old child who sees one for the first time in a storybook. A zoologist may have a complex schema composed of physical, behavioral, historical, ecological, and perhaps other elements. The child's elephant schema may be connected only to the other characters of the story. The zoologist will have connections between the elephant and many schema throughout his or her long-term memory network.

Another important characteristic of memory is the strength of the connections between the units. Memory strength is developed through repetition, salience, and/or elaboration. Each time a memory is experienced, the ease with which that memory can be recalled in the future increases (Hebb, 1976). Thus, rote memorization increases memory strength by increasing the number of times the memory was activated. Similarly, experiences that have strong sensory or emotional elements have a disproportionate gain in memory strength. A workplace error that has significant consequences will be remembered much better than one that has none. Elaboration involves relating the new information to existing schema and incorporating it in an organized way. Memory strength has a substantial impact on cognition. Well-learned schema can be recalled faster and with less effort because less energy is required to activate the stronger connections.

Types of Long-Term Memory

Long-term memory can be partitioned into categories such as episodic, semantic, and procedural components (Tulving, 1989).

Episodic memory refers to traces that remain from the individual's personal experiences. Events from the past are stored as subsensory cell assemblies that are connected to maintain important features of the event but generalize less important features. Thus, an athlete's episodic memory of the championship game may include specific and detailed visual traces of significant actions during the game. But less important actions may really be represented using statistical aggregates of similar actions that occurred over the athlete's total experience of games. These aggregates would be reconstructed during recall to provide the recall experience.

Semantic memory is composed of conceptual information, including knowledge of concept definitions, object relationships, physical laws, and similar nonsensory information (Lachman, Lachman, and Butterfield, 1979). Connections within semantic memory link concepts that are related or are used together, while the composition of semantic memory is not as structured or organized as an explicit semantic network. This is a reasonable simplification for the purposes of this chapter. There are also links between semantic memory and episodic memory. For example, the semantic memory of the meaning of "quality" may be linked to the episodic memory of a high-quality product.

Procedural memory refers to the combination of muscle movements that compose a unitary action (Johnson, 2003). Procedural memories are often unconscious and stored as one complete unit. For example, a pianist may be able to play a complex piano concerto but cannot verbally report what the fourth note is without first imagining the initial three notes. These automatic processes can take over in emergency situations when there is not enough time to think consciously about required actions.

Memories are not separated into distinct units that are clearly demarcated within the human information-processing system. There are overlaps and interconnections within the memory structure that have both advantages and disadvantages for human performance. Creative problem-solving is enhanced when rarely used connections are activated to brainstorm for solution ideas.

But this can lead to errors when random connections are assumed to represent fact or statistical associations are assumed to apply to inappropriate specific cases.

Implications for Design

The structure of long-term memory has significant implications for the design of industrial systems. Workers can master work activities faster when new processes match prior learning and overlap with existing schema. This will also reduce the probability that inappropriate schema will become activated in emergency situations, perhaps leading to errors. Similarly, terminology for labels, instructions, and information displays should be unambiguous to maximize speed of processing.

Simulator and field training can focus on developing episodic memories that bolster semantic memory for system architecture and underlying physical laws. Training breadth can expand the semantic schema to include a variety of possible scenarios, while repetition can be used to solidify the connections for critical and frequent activities. Scielzo, Fiore, Cuevas, and Salas (2002) found that training protocols that supported the development of accurate schema allowed new learners' performance to approach that of experts.

Case: Power-Plant Control

Control rooms are often composed of a large set of monitors, controls, and displays that show the status of processes in graphical, tabular, and digital readouts. Operators are trained to recognize problems as soon as they occur, diagnose the problem, and initiate steps to correct the problem. The design of this training is critical so that operators develop schema that effectively support problem-solving.

To maximize the ability of operators to identify major emergencies, training should include repeated simulations of these emergencies. According to Wickens, Lee, Liu, and Gordon-Becker (2004), initial training leads to an accurate response for a given situation. But additional training is still necessary to increase the speed of the response and to reduce the amount of attention that is necessary for the emergency to be noticed and recognized. Thus, overlearning is important for emergency response tasks. With sufficient repetition, operators will have strong long-term memories for each emergency and can recognize them more quickly and accurately (Klein, 1993). They will know which combination of displays will be affected for each problem and what steps to take. Accurate feedback is critical to ensure that workers associate the correct responses with each emergency (Wickens et al., 2004). When errors are made, corrective feedback is particularly important so that the error does not become part of the learned response.

But problems do not always occur in the same way. In order for training to cover the variability in problem appearance, variation must be included in the training. Employees must be trained to recognize the diversity of situations that can occur. This will develop broader schema that reflect a deeper conceptual understanding of the problem states and lead to a better ability to implement solutions. Semantic knowledge is also important because procedures can be context specific (Gordon, 1994). Semantic knowledge helps employees to adapt existing procedures to new situations.

Training fidelity is also an important consideration. The ecological validity of training environments has been shown to increase training transfer, but Swezey and Llaneras (1997) have shown that not all features of the real environment are necessary. Training design should include an evaluation of what aspects of the real environment contribute to the development of effective problem schema.

Working Memory
While the entire network of schema stored in long-term memory is extensive, it is impossible for an individual to recall simultaneously more than a limited set. Working memory refers to the set of schema that is activated at one point in time. A schema stored in long-term memory reverberates due to some input stimulus and can remain activated even when the stimulus is removed (Jones

and Polk, 2002). Working memory can consist of a combination of semantic, episodic, and procedural memories. It can be a list of unrelated items (such as a shopping list), or it can be a situational model of a complex environment (see Situation Awareness section later in this chapter).

The size of working memory has been the focus of a great body of research. Miller's (1956) famous study that reported a span of 7 ± 2 is widely cited. However, this is an oversimplification. It depends on the size and complexity of the schema being activated. Many simple schemas (such as the single digits and letters used in much of the original psychology research) can be more easily maintained in working memory compared to complex schema (such as mental models of system architecture), because the amount of energy required to activate a complex schema is greater (see the section Attention and Mental Workload later in this chapter). Experience is also a factor. Someone who is an expert in aerospace systems has stronger aerospace systems schema and, therefore, can recall schema related to aerospace systems with less effort because of this greater strength. More schemas can thus be maintained in working memory.

The size of working memory also depends on the ability of the worker to combine information into chunks (Wickens et al., 2004). Chunks are sets of working memory units that are combined into single units based on perceptual or semantic similarity. For example, mnemonics enhance memory by allowing workers to remember a single acronym, such as SEARCH for simplify, eliminate, alter sequence, requirements, combine operations, and how often for process improvement brainstorming (from Konz and Johnson, 2000), more easily than a list of items.

How long information can be retained in working memory depends on the opportunity for workers to subvocally rehearse the information (Wickens et al., 2004). Without rehearsal, information in working memory is lost rapidly. Thus, when working memory must be heavily used, distractions must be minimized, and ancillary tasks that also draw on this subvocalization resource must be avoided.

The similarity of competing information in working memory also affects the reliability of recall. Because working memory exists in the auditory modality, information that sounds alike is most likely to be confused (Wickens et al., 2004). The working memory requirements for any concurrent activities must be considered to minimize the risk of interference.

Implications for Design

A better understanding of working memory can support the development of more reliable industrial systems. The amount of information required to complete work activities should be considered in relation to the working memory capacities of the workers. Norman (1988) describes two categories of information storage. Information in the head refers to memory, and information in the world refers to labels, instructions, displays, and other physical devices. When the amount of information to complete a task exceeds the capacity of working memory, it must be made available in the physical world, through computer displays, manuals, labels, and help systems. Of course, accessing information in the world takes longer than recalling it from memory, so this time must be considered when evaluating system performance requirements. This trade-off can also affect accuracy, as workers may be tempted to use unreliable information in working memory to avoid having to search through manuals or displays for the correct information (Gray and Fu, 2004).

Similarly, when information must be maintained in working memory for a long period of time, the intensity can fall below the threshold required for reliable recall. Here too, important information should be placed in the physical world. Interfaces that allow workers to store preliminary hypotheses and rough ideas can alleviate the working memory requirements and reduce the risk of memory-related errors. When information must be maintained in working memory for extended periods, the worker must be allowed to focus on rehearsal. Any other tasks that require the use of working memory must be avoided. Distractions that can interfere with working memory must be eliminated.

Training can also be used to enhance working memory. Training modules can be used to strengthen workers' conception of complex processes, and thus can reduce the working memory required to maintain it during work activities. This would allow additional information to be considered in parallel.

Case: Cockpit Checklists
Degani and Wiener (1993) describe cockpit checklists as a way to provide redundancy in configuring an aircraft and reduce the risk of missing a step in the configuration process. Without checklists, aircraft crews would have to retrieve dozens of configuration steps from long-term memory and maintain in working memory whether each step had been completed for the current flight. Checklists reduce this memory load by transferring the information into the world. Especially in environments with frequent interruptions and distractions, the physical embodiment of a procedure can ensure that no steps are omitted because of lapses in working memory.

Sensation

Sensation is the process through which information about the world is transferred to the brain's perceptual system through sensory organs, such as the eyes and ears (Bailey, 1996). From a systems design perspective, there are three important parameters for each dimension that must be considered: sensory threshold, difference threshold, and stimulus–response ratio.

The sensory threshold is the level of stimulus intensity below which the signal cannot be sensed reliably. The threshold must be considered in relation to the work environment. In the visual modality, there are several important stimulus thresholds. For example, in systems that use lights as warnings, indicators and displays need to have a size and brightness that can be seen by workers at the appropriate distance. These thresholds were determined in ideal environments. When environments are degraded because of dust or smoke, or workers are concentrating on other tasks, the thresholds may be much higher. In environments with glare or airborne contaminants, the visual requirements will change.

Auditory signals must have a frequency and intensity that workers can hear, again at the appropriate distance. Workplaces that are loud or where workers will be wearing hearing protection must be considered. Olfactory, vestibular, gustatory, and kinesthetic senses have similar threshold requirements that must be considered in system design.

The difference threshold is the minimum change in stimulus intensity that can differentiated; this is also called the "just noticeable difference" (Snodgrass, Levy-Berger, and Haydon, 1985). This difference is expressed as a percent change. For example, a light must be at least 1 percent brighter than a comparison for a person to be able to tell that they are different. A sound must be 20 percent louder than a comparison for a person to perceive the difference.

The difference threshold is critical for the design of systems when multiple signals must be differentiated. When different alarms are used to signal different events, it is critical that workers be able to recognize the difference. When different-sized connectors are used for different parts of an assembly, workers need to be able to distinguish which connector is the correct one. Although there has been little research in this area, it is likely that there is a speed/accuracy trade-off with respect to difference thresholds. When workers are forced to act quickly, either because of productivity standards or in an emergency situation, even higher differences may be required for accurate selection.

The third dimension is the stimulus–response ratio. The relationship between the increase in intensity in a sensory stimulus and the corresponding increase in the sensation of that intensity is an exponential function (Stevens, 1975). For example, the exponent for perception of load heaviness is 1.45, so a load that is 1.61 times as heavy as another load will be perceived as twice as heavy. Similarly, the exponent for the brightness of a light is 0.33, so a light has to be eight times as bright to be perceived as twice as bright. Predicting these differences in perception is critical so that systems can be designed to minimize human error in identifying and responding to events.

Implications for Design

To maximize the reliability with which important information will reach workers, work environment design must consider sensation. The work environment must be designed to maximize the clarity with which workers can sense important sources of information. Lighting must be maintained to allow workers to see at requisite accuracy levels. Effective choice of color for signs and displays can maximize contrast with backgrounds and the accuracy of interpretation. Background noise can be controlled to allow workers to hear important signals and maintain verbal communication. The frequency and loudness of auditory signals and warnings can be selected to maximize comprehension. Location is also important. Key sources of visual information should be placed within the worker's natural line of sight.

Case: Industrial Dashboards

Designing system interfaces to support complex decision-making, such as with supply chain, enterprise, and executive dashboards, requires a focus on human sensory capabilities (Resnick, 2003). Display design requires selecting among digital, analog, historical, and other display types (Hansen, 1995; Overbye, Sun, Wiegmann, and Rich, 2002). The optimal design depends on how often the data change and how quickly they must be read.

The salience of each interface unit is also critical to ensure that the relevant ones attract attention from among the many others on the display (Bennett and Flach, 1992). A variety of techniques can be used in industrial dashboards to create salience, such as brightness, size, auditory signals, or visual animation. The design should depend on the kinds of hardware on which the system will be implemented. For example, when systems will be accessed through handheld or notebook computers, the display size and color capabilities will be limited, and these limitations must be considered in the display design.

Perception

As these basic sensory dimensions are resolved, the perceptual system tries to put them together into identifiable units. This is where each sensation is assigned to either an object or the background. If there are several objects, the sensations that compose each of them must be separated. There is a strong interaction here with long-term memory (see section entitled Long-Term Memory). Objects with strong long-term memory representations can be recognized faster and more reliably because less energy is required to activate the corresponding schema. Objects that have similar features to different but well-known objects are easily misidentified as these objects. This is called a *capture error* because of the way the stronger schema "captures" the perception and becomes active first.

There is also an interaction with working memory (see section Attention and Mental Workload). Objects that are expected to appear are also recognized faster and more reliably. Expectations can be described as the priming of the schema for the object that is expected. Energy is introduced into the schema before the object is perceived. Thus, less actual physical evidence is needed for this schema to reach its activation threshold. This can lead to errors when the experienced object is not the one that was expected but has some similarities.

Implications for Design

The implications of perception for industrial systems design are clear. When designing work objects, processes, and situations, there is a trade-off between the costs and benefits of similarity and overlap. When it is important that workers are able to distinguish objects immediately, particularly in emergency situations, overlap should be minimized. Design efforts should focus on the attributes that workers primarily use to distinguish similar objects. Workers can be trained to focus on features that are different. When object similarity cannot be eliminated, workers can be trained to recognize subtle differences that reliably denote the object's identity.

It is also important to control workers' expectations. Because expectations can influence object recognition, it is important that they reflect the true likelihood of object presence. This can be accomplished through situational training. If workers know what to expect, they can more quickly and accurately recognize objects when they appear. For those situations where there is too much variability for expectations to be reliable, work procedures can include explicit re-checking of the identity of objects where correct identification is critical.

Case: In-Vehicle Navigation Systems
In-vehicle navigation systems help drivers find their way by showing information on how to travel to a programmed destination. These systems can vary greatly in the types of information that they provide and the ways in which the information is presented. For example, current systems can present turn-by-turn directions in the visual and/or auditory modalities, often adjusted according to real-time traffic information. These systems can also show maps that have the recommended route and traffic congestion highlighted in different colors.

There are many advantages provided by these systems. In a delivery application, optimization software can consider all of the driver's remaining deliveries and current traffic congestion to compute the optimal order to deliver the packages. For many multistop routes, this computation would exceed the driver's ability to process the information. Including real-time traffic information also enhances the capabilities of the system to select the optimal route to the next destination (Khattak, Kanafani, and Le Colletter, 1994).

A challenge for these systems is to provide this information in a format that can be quickly perceived by the driver. Otherwise, there is a risk that the time required for the driver to perceive the relevant information will require an extended gaze duration, increasing the likelihood of a traffic accident. Persaud and Resnick (2001) found that the display modality had a significant effect on decision-making time. Graphical displays, although the most common design, required the most time to parse. Recall scores were also lowest for graphical displays, possibly requiring the driver to look back at the display more often. This decrease in recognition speed can lead to greater risk of a traffic accident.

Attention and Mental Workload

Because only a small subset of long-term memory can be activated at any one time, it is important to consider how this subset is determined. Ideally, attention will be focused on the most important activities and the most relevant components of each activity, but the prevalence of errors that are the result of inappropriately focused attention clearly indicates that this is not always the case.

There are many channels of information, both internal and external, on which attention can be focused. In most industrial settings, there can be visual and auditory displays that are designed specifically to present information to workers to direct their job activities. There are also informal channels in the various sights, sounds, smells, vibrations, and other sensory emanations around the workplace. Communication with other workers is also a common source of information. Additionally, there are internal sources of information in the memory of the individual. Episodic and semantic memories both can be a focus of attention. But it is impossible for workers to focus their attention on all of these channels at once.

It is also important to consider that attention can be drawn to channels that are relevant to the intended activities, but also to those that are irrelevant. Daydreaming is a common example of attention being focused on unessential information channels. Attention is driven in large part by the salience of each existing information channel. Salience can be defined as the attention-attracting properties of an object. It can be derived based on the intensity of a channel's output in various sensory modalities (Wickens and Hollands, 2000). For example, a loud alarm is more likely to draw attention than a quiet alarm. Salience can also be based on the semantic importance of the channel. An alarm that indicates a nuclear accident is more likely to draw attention than an alarm signaling lunchtime. Salience is the reason that workers tend to daydream when work intensity is low, such as long-duration monitoring of displays (control center operators, air travel, and security). When nothing is happening on the display, daydreams are more interesting and draw away the worker's attention. If something important happens, the worker may not notice.

If humans had unlimited attention, then we could focus on all possible information sources, internal and external. However, this is not the case; there is a limited amount of attention available. The number of channels on which attention can be focused depends on the complexity of each channel. One complex channel, such as a multifunction display, may require the same amount of attention as several simple channels, such as warning indicators.

Another important consideration is the total amount of attention that is focused on an activity and how this amount varies over time. This *mental workload* can be used to measure how busy a worker is at any given time, to determine if any additional tasks can be assigned without degrading performance, and to predict whether a worker could respond to unexpected events. Mental workload can be measured in several ways, including the use of subjective scales rated by the individual doing the activity or physiologically by measuring the individual's heart rate and/or brain function. A great deal of research has shown that mental workload must be maintained within the worker's capability, or job performance will suffer in domains such as air-traffic control (Lamoreux, 1997), driving a car (Hancock et al., 1990), and others.

Implications for Design

There are many ways to design the work environment to facilitate the ability of the worker to pay attention to the most appropriate information sources. Channels that are rarely diagnostic should be designed to have low salience. Important channels can be designed to have high sensory salience through bright colors or loud auditory signals. Salient auditory alerts can be used to direct workers' attention toward key visual channels. Workers should be trained to recognize diagnostic channels so that they evoke high semantic salience.

Mental workload should also be considered in systems design. Activities should be investigated to ensure that peak levels of workload are within workers' capabilities. Average workload should not be so high as to create cumulative mental fatigue. It should also not be so low that workers are bored and may miss important signals when they do occur.

Case: Warnings

A warning is more than a sign conveying specific safety information. It is any communication that reduces risk by influencing behavior (Laughery and Hammond, 1999). One of the most overlooked aspects of warning design is the importance of attention. In a structured recall environment, a worker may be able to accurately recall the contents of a warning. However, if a warning is not encountered during the activity in which it is needed, it may not affect behavior because the worker may not think of it at the time when it is needed. When the worker is focusing on required work activities, the contents of the warning may not be sufficiently salient to direct safe behavior (Wogalter and Leonard, 1999). Attention can be attracted with salient designs, such as bright lights, sharp contrasts, auditory signals, large sizes, and other visualization enhancements.

Frantz and Rhoades (1993) reported that placing warnings in locations that physically interfered with the task could increase compliance even further. The key is to ensure that the warning is part of the attentional focus of the employee at the time it is needed and that it does not increase mental workload past the employees' capacity.

Situation Awareness

Situation awareness (SA) is essentially a state in which an observer understands what is going on in his or her environment (Endsley, 2000a). There are three levels of SA: perception, comprehension, and projection. Perceptional SA requires that the observer know to which information sources attention should be focused and how to perceive these sources. In most complex environments, many information sources can draw attention. Dividing one's attention among all of them reduces the time that one can spend on critical cues and increases the chance that one may miss important events. This is Level 1 SA, which can be lacking when relevant information sources have low salience, when they are physically obstructed, when they are not available at needed times, when there are distractions, or when the observer lacks an adequate sampling strategy (Eurocontrol, 2003). The observer must be able to distinguish three types of information sources: those that must be examined and updated constantly, those that can be searched only when needed, and those that can be ignored.

Comprehension is the process of integrating the relevant information that is received into a cohesive understanding of the environment and retaining this understanding in memory for as long as it is needed. This includes both objective analysis and subjective interpretation (Flach, 1995). Comprehension can be compromised when the observer has an inadequate schema of the work environment or is over-reliant on default information or default responses (Eurocontrol, 2003).

Projection is when an observer can anticipate how the situation will evolve over time, can anticipate future events, and comprehends the implications of these changes. The ability to project supports timely and effective decision-making (Endsley, 2000a). One key aspect of projection is the ability to predict *when* and *where* events will occur. Projection errors can occur when current trends are under- or overprojected (Eurocontrol, 2003).

Endsley (2000a) cautions that SA does not develop only from official system interface sources. Workers can garner information from informal communication, world knowledge, and other unintended sources. SA is also limited by attention demands. When mental workload exceeds the observer's capacity, either because of an unexpected increase in the flow of information or because of incremental mental fatigue, SA will decline.

Implications for Design

Designing systems to maximize situation awareness relies on a comprehensive task analysis. Designers should understand each goal of the work activity, the decisions that will be required, and the best diagnostic information sources for these decisions (Endsley, 2002).

It is critical to predict the data needs of the worker in order to ensure that these are available when they are needed (Endsley, 2001). However, overload is also a risk, because data must be absorbed and assimilated in the time available. To avoid overload, designers can focus on creating information sources that perform some of the analysis in advance and present integrated results to the worker. Displays can also be goal-oriented, and information can be hidden at times when it is not needed.

It is also possible to design set sequences for the sampling of information sources into the work processes. This can enhance situation awareness because the mental workload due to task overhead is reduced. Workers should also be informed of the diagnosticity of each information source.

Case: Air-Traffic Systems

Situation awareness has been used in the investigation of air-traffic incidents and to identify design air-traffic control modifications that can reduce the likelihood of future incidents (Rodgers, Mogford and Strauch, 2000). In the latter study, inadequate SA was linked to poor decision-making quality, leading to both minor incidents and major aircraft accidents. When air-traffic controllers are aware of developing error situations, the severity of the incident is reduced. The study identified several hypotheses to explain the loss of SA in both high-workload and low-workload situations. In high-workload conditions, operators had difficulty maintaining a mental picture of the air traffic. As the workload shifts down from high to low, sustained periods can lead to fatigue-induced loss of SA. The evaluation of air-traffic controller situation awareness led to insights into the design of the radar display, communication systems, team coordination protocols, and data-entry requirements.

Situation awareness has also been used in the design stage to evaluate competing design alternatives. For example, Endsley (2000b) compared sensor hardware, avionics systems, free flight implementations, and levels of automation for pilots. These tests were sensitive, reliable, and able to predict the design alternative that achieved the best performance.

Decision-Making

Decision-making is the process of selecting an option based on a set of information under conditions of uncertainty (Wickens et al., 2004). Contrary to the systematic way that deliberate decisions are made or programmed into computers, human decision-making is often unconscious, and the specific mechanisms are unavailable for contemplation or analysis by the person who made them. Environments with many interacting components, degrees of freedom, and/or unclear data sources challenge the decision-making process. Decision-making processes are affected by neurophysiological characteristics that are influenced by the structure of long-term memory and the psychological environment in which the decision is made. For experienced decision-makers, decisions are situational discriminations (Dreyfus, 1997) where the answer is obvious without comparison of alternatives (Klein, 2000). There are two major types of decision-making situations: diagnosis and choice.

Diagnosis

Diagnosis decisions involve evaluating a situation to understand its nature, and can be modeled as a pattern recognition process (Klein, 2000). Diagnosis describes decisions made in troubleshooting, medical diagnosis, accident investigation, safety behavior, and many other domains. The information that is available about the situation is compared to the existing schema in long-term memory, subject to the biasing effects of expectations in working memory. If there is a match, the corresponding schema becomes the diagnosis. For experts, this matching process can be modeled as a recognition-primed decision (Klein, 1993), whereby the environment is recognized as matching one particular pattern, and the corresponding action is implemented.

The minimum degree to which the current situation must match an existing schema depends on the importance of the decision, the consequences of error, and the amount of time available. When the cost of searching for more information exceeds the expected benefits of that information, the search process stops (Marble, Medema, and Hill, 2002). For important decisions, this match threshold will be higher so that more evidence can be sought before a decision is made. This leads to more reliable and accurate decisions. However, it may still be the case that an observed pattern matches an existing schema immediately, and a decision is made regardless of how important the decision may be.

Under conditions of time pressure, there may not be sufficient time to sample enough information channels to reach the appropriate threshold. In these cases, the threshold must be lowered, and decisions will be made based only on the information available (Ordonez and Benson, 1997). In these cases, individuals focus on the most salient source of information (Wickens and Hollands, 2000) and select the closest match based on whatever evidence has been collected at that point (Klein, 1993).

When the decision-maker is an expert in the domain, this process is largely unconscious. The matched schema may be immediately apparent with no one-by-one evaluation of alternatives. Novices may have less well-structured schema, and so the match will not be clear. More explicit evaluation may be required.

Choice

In choice decisions, an individual chooses from a set of options that differ in the degree to which they satisfy competing goals. For example, when one is choosing a car, one model may have a better safety record and another may be less expensive. Neither is necessarily incorrect, although one may be more appropriate according to a specific set of optimization criteria.

When a person makes a decision, it is often based on an unconscious hybrid of several decision-making strategies (Campbell and Bolton, 2003). In the weighted average strategy, the score on each attribute is multiplied by the importance of the attribute, and the option with the highest total score is selected (Jedetski, Adelman, and Yeo, 2002). However, this strategy generally requires too much information processing for most situations and often does not match the desired solution (Campbell and Bolton, 2003). In the satisficing strategy, a minimum score is set for each attribute. The first option that meets all of these minima is selected (Simon, 1955). If none do, then the minimum of the least important attribute is relaxed and so on until an option is acceptable. In the lexicographic strategy, the option with the highest score on the most important attribute is selected without regard for other attributes (Campbell and Bolton, 2003).

Using the weighted adding strategy, option 1 would receive 173 points ($7 \times 8 + 3 \times 5 + 8 \times 9 + 5 \times 6$). Options 2 and 3 would receive 168 and 142, respectively. So, option 1 would be selected. The company may have satisficing constraints for attributes such as safety and value. A safety score less than 5 and a value score below 4 may be considered unacceptable regardless of the other attribute scores (eliminating options 1 and 2 from consideration), resulting in the selection of option 3. Finally, the company may choose to use a lexicographic strategy on value, selecting the option with the highest value regardless of all other attribute scores. In this case, option 2 would be selected.

While the weighted adding strategy is often considered the most optimal, this is not necessarily the case. Some attributes, such as safety, should not be compensatory. Regardless of how fast, capable, reliable, or cost-effective a machine may be, risk to workers' safety should not be compromised. Lexicographic strategies may be justified when one attribute dominates the others, or the company does not have the time or resources to evaluate other attributes. For example, in an emergency situation, preventing the loss of life may dominate consideration of cost or equipment damage. The use of these strategies can be quite effective (Giggerenzer and Todd, 1999). And according to Schwartz (2004), benefits gained from making optimal decisions are often not worth the time and effort required.

Contrary to the systematic way the companies make official decisions, day-to-day decisions are often made with little conscious evaluation of the strategy (reference). As with diagnosis decisions, time pressure and decision importance influence the decision-making process. When faced with limited time, workers may be forced to use faster, simpler strategies such as the lexicographic strategy (Ordonez and Benson, 1997).

Decision-Making Heuristics

There are several decision-making heuristics that can reduce the information-processing requirements and often reduce the time required to make a decision. However, these shortcuts can also bias the eventual outcome (Browne and Ramesh, 2002). These are often not consciously applied, so they can be difficult to overcome when they degrade decision-making accuracy and reliability.

- *Anchoring*: When an individual develops an initial hypothesis in either a diagnosis or choice decision, it is very difficult to switch to an alternative. Contrary evidence may be discounted.

- *Confirmation*: When an individual develops an initial hypothesis in either a diagnosis or choice decision, he or she will have a tendency to search for information that supports this hypothesis even when other channels may be more diagnostic.

- *Availability*: When searching for additional information, sources that are more easily accessed or brought to mind will be considered first, even when other sources are more diagnostic.

- *Reliability*: The reliability of information sources is hard to integrate into the decision-making process. Differences in reliability are often ignored or discounted.

- *Memory limitations*: Because of the higher mental workload required to keep many information sources in working memory simultaneously, the decision-making process will often be confined to a limited number of information sources, hypotheses, and attributes.

- *Feedback*: Similar to the confirmation bias, decision-makers often focus on feedback that supports a past decision and discount feedback that contradicts past decisions.

Implications for Design

Human factors can have a tremendous impact on the accuracy of decision-making. It is often assumed that normative decision-making strategies are optimal and that workers will use them when possible. However, neither of these is the case in many human decision-making situations. Limitations in information-processing capability often force workers to use heuristics and focus on a reduced number of information sources. Competing and vague goals can reduce the applicability of normative decision criteria.

Workers can be trained to focus on the most diagnostic sources in each decision domain. If they are only going to use a limited number of sources, they should at least be using the most effective ones. Diagnostic sources also can be given prominent locations in displays or be the focus of established procedures.

The reliability of various information sources should be clearly visible either during the decision-making process or during training. Workers can be trained to recognize source reliability or to verify it in real time. Similarly, workers can be trained to recognize the best sources of feedback. In design, feedback can be given a more prominent position or provided more quickly.

To avoid anchoring and confirmation biases, decision support systems can be included that suggest (or require) workers to consider alternatives, seek information from all information sources, and include these sources in the decision-making process. At the least, a system for workers to externalize their hypotheses will increase the chance that they recognize these biases when they occur. However, the most successful expert systems are those that complement the human decision process rather than stand-alone advisors that replace humans (Roth, Bennett, and Woods, 1987).

In cases where decision criteria are established in advance, systems can be designed to support the most effective strategies. Where minimum levels of performance for particular criteria are important, the decision support systems can assist the worker in establishing the level and eliminating options that do not reach this threshold. The information-processing requirements of weighted adding strategies can be offloaded to decision support systems entirely to free the worker for information-collecting tasks for which he or she may be more suited.

Case: Accident Investigation

Accident investigation and the associated root cause analysis can be fraught with decision-making challenges. Human error is often the proximate cause (Mullen, 2004) of accidents but is much less often the root cause. During the accident investigation process, it is critical for investigators to explore the factors that led to the error and identify the design changes that will eliminate future risk (Doggett, 2004). However, this process engenders many opportunities for decision-making errors. Availability is usually the first obstacle. When an accident occurs, there is often high-visibility evidence that may or may not lead directly to the root cause. The CAIB report found that the root cause that ultimately led to the Columbia accident was not a technical error related to the foam shielding that was the early focus of the investigation, but rather was due to the organizational culture of NASA (CAIB, 2003). The confirmation bias can also challenge the investigation process. When investigators develop an initial hypothesis that a particular system component led to an accident, they may focus exclusively on evidence to confirm this component as the root cause rather than general criteria that could rule out other likely causes. This appeared in the investigations of the USS *Vincennes* incident in the Persian Gulf and the Three Mile Island nuclear power incident. Decision support systems can be used to remove a lot of the bias and assist in the pursuit of root causes. By creating a structure around the investigation, they can lead investigators to diagnostic criteria and ensure that factors such as base rates are considered. Roth, Gualtieri, Elm, and Potter (2002) provide an overview of how decision support systems (DSSs) can reduce bias in decision-making. For example, DSSs can inform users when the value for a particular piece of evidence falls outside a specified range. They can make confirming and disconfirming directions explicit and facilitate switching between them. But they warn that these systems can also introduce errors, such as by allowing drill-down into large data sources so that many data in one area are sampled without looking elsewhere. DSSs can also exacerbate the availability bias by providing easy access to recent investigations.

COGNITIVE CONSEQUENCES OF DESIGN

Learning

Every time an object is perceived, an event is experienced, a memory is recalled, or a decision is made, there are small, incremental changes in the structure of the human information-processing system. Learning is very difficult when there is no prior experience to provide a framework (Hebb, 1955). This explains the power of analogies in early training. Later learning is a recombination of familiar patterns through the transition of general rules into automatic procedures and the generalization and specialization of these procedures as experience develops (Taatgen and Lee, 2003). The magnitude of the change depends on the salience of the experience and how well it matches existing schema.

When a human-system interaction is the same as past experiences, there is very little learning because no new information is gained. The only result is a small strengthening of the existing schema. It is unlikely that the worker will develop a strong episodic memory of the event. When a human-system interaction is radically different from anything that has been experienced before, a strong episodic memory may be created because of the inherent salience of confusion and possible danger. But the event will not be integrated into the semantic network because it does not correspond to any of the existing schema—there is nowhere to "put" it. Maximum learning occurs when a human-system interaction mirrors past experience but has new attributes that make sense; that is, the experience can be integrated into the conceptual understanding of the system.

Implications for Design

Training programs should always be designed based on an analysis of the workers' existing knowledge. Training of rote procedures where there will be no variability in workers' actions should be approached differently than training for situations where workers will be required to recognize and solve problems. In a study of novice pilot training, Fiore, Cuevas, and Oser (2003) found that diagrams and analogical representations of text content facilitate learning of procedures that require knowledge elaboration, but not on recognition or declarative rote memorization.

A better understanding of human learning mechanisms can also facilitate the development of experiential learning that workers gain on the job. System interfaces can be structured to maximize experiential learning by providing details that help employees develop accurate schema of the problem space. Over time, repeated exposure to this information can lead to more detailed and complex schema that can facilitate more elaborate problem-solving. A cognitive analysis of the task requirements and possible situations can lead to a human-system interface that promotes long-term learning.

Error

Human behavior is often divided into three categories: skill based, rule based, and knowledge based (Rasmussen, 1993). In skill-based behavior, familiar situations automatically induce well-practiced responses with very little attention. In deterministic situations with a known set of effective responses, simple IF-THEN decision criteria lead to rule-based behaviors. Knowledge-based behaviors are required in unfamiliar or uncertain environments where problem-solving and mental simulation are required.

Each of these behavior types is associated with different kinds of errors (Reason, 1990). With skill-based behavior, the most common errors are related to competing response schema. Skill-based behaviors result from a strong schema that is associated repeatedly with the same response. When a new situation shares key attributes with this strong schema, the old response may be activated in error. Because skill-based behaviors require little attention, the response is often completed before the error is noticed. In these cases, expertise can actually hurt performance accuracy. Unless there is salient feedback, the error may not be noticed, and there will be no near-term recovery from the error.

Rule-based behaviors can lead to error when a rule is erroneously applied, either because the situation was incorrectly recognized or because the rule is inappropriately generalized to similar situations. Rule-based behavior is common with novices who are attempting to apply principles acquired in training. Because rule-based behavior involves conscious attention, the error is likely to be noticed, but the employee may not know of a correct response to implement.

Knowledge-based errors occur when the employee's knowledge is insufficient to solve a problem. Knowledge-based behavior is the most likely to result in error, because it is the type of behavior most often used in uncertain environments. When an employee is aware that his or her schema is not sophisticated enough to predict how a system will respond, he or she may anticipate a high likelihood of error and specifically look for one. This increases the chance that errors will be noticed and addressed.

Implications for Design

If system designers can anticipate the type(s) of behavior that are likely to be used with each employee–system interaction, steps can be taken to minimize the probability and severity of errors that can occur. For example, when skill-based behavior is anticipated, salient feedback must be designed into the display interface to ensure that employees will be aware when an error is made. To prevent skill-based errors from occurring, designers can make key attributes salient so that the inappropriate response will not be initiated.

To prevent rule-based errors, designers should ensure that the rules taught during training match the situations that employees will encounter when they are interacting with the system later. The triggers that indicate when to apply each rule should be made explicit in the system interface design. Signals that indicate when existing rules are not appropriate should be integrated into the interface design.

For complex systems or troubleshooting scenarios, when knowledge-based behavior is likely, errors can only be minimized when employees develop effective schema of system operators or when problem-solving activities are supported by comprehensive documentation and/or expert systems. Training should ensure that employees are aware of what they know and what they do not know. Employee actions should be easy to reverse when they are found to be incorrect.

SUMMARY

Humans interact with industrial systems throughout the system life cycle. By integrating Human Factors into each stage, the effectiveness, quality, reliability, efficiency, and usability of the system can be enhanced. At the requirements stage, it is critical for management to appreciate the complexity of human-system interaction and allocate sufficient resources to ensure that Human Factors requirements are emphasized. During design, Human Factors should be considered with the earliest design concepts to maximize the match between human capabilities and system operations. As the system develops, Human Factors must be applied to control and display design and the development of instruction and training programs. Maintenance operations should also consider Human Factors to ensure that systems can be preserved and repaired effectively. Human Factors is also critical for human error analysis and accident investigation. This chapter presented a model of human information processing that addresses most of the relevant components of human cognition. Of course, one chapter is not sufficient to communicate all of the relevant Human Factors concepts that relate to the system life cycle. But it does provide a starting point for including Human Factors in the process.

In addition to describing the critical components of human cognition, this chapter described some of the implications of human cognition on system design. These guidelines can be applied throughout systems design. The specific cases are intended to illustrate this implementation in a variety of domains. As technology advances and the nature of human-system interaction changes, research will be needed to investigate specific human-system interaction effects. But an understanding of the fundamental nature of human cognition and its implications for system performance can be a useful tool for the design and operation of systems in any domain.

REFERENCES

Bailey, R. W. 1996, *Human Performance Engineering*, 3rd ed. Upper Saddle River, NJ: Prentice Hall.

Bennett, K. B. and Flach, J. M. 1992, Graphical Displays: Implications for Divided Attention, Focused Attention, and Problem Solving, *Human Factors*, Vol. 34, No. 5, pp. 513–533.

Browne, G. J. and Ramesh, V. 2002, Improving Information Requirements Determination: A Cognitive Perspective, *Information and Management*, Vol. 39, pp. 625–645.

CAIB. 2003, *The Columbia Accident Investigation Board Final Report*, NASA. Retrieved August 4, 2004, at http://www.caib.us.

Campbell, G. E. and Bolton, A. E. 2003, Fitting Human Data with Fast, Frugal, and Computable Models of Decision Making, in *Proceedings of the Human Factors and Ergonomics Society 47th Annual Meeting*, Human Factors and Ergonomics Society, Santa Monica, CA, pp. 325–329.

Degani, A. and Wiener, E. L. 1993, Cockpit Checklists: Concepts, Design, and Use, *Human Factors*, Vol. 35, No. 2, pp. 345–359.

Doggett, A. M. 2004, A Statistical Comparison of Three Root Cause Analysis Tools, *Journal of Industrial Technology*, Vol. 20, No. 2, pp. 2–9.

Dreyfus, H. L. 1997, Intuitive, Deliberative, and Calculative Models of Expert Performance, in Zsambok, C. E. and Klein, G. (Eds.), *Naturalistic Decision Making*. Mahwah, NJ: Lawrence Erlbaum.

Endsley, M. R. 2000a, Theoretical Underpinnings of Situation Awareness: A Critical Review, in Endsley, M. R. and Garland, D. J. (Eds.), *Situation Awareness Analysis and Measurement*. Mahwah, NJ: Lawrence Erlbaum.

Endsley, M. R. 2000b, Direct Measurement of Situation Awareness: Validity and Use of SAGAT, in Endsley, M. R. and Garland, D. J. (Eds.), *Situation Awareness Analysis and Measurement*. Mahwah, NJ: Lawrence Erlbaum.

Endsley, M. R. 2001, Designing for Situation Awareness in Complex Systems, in *Proceedings of the Second International Workshop of Symbiosis of Humans, Artifacts, and Environments*, Kyoto, Japan.

Endsley, M. R. 2002, From Cognitive Task Analysis to System Design, CTAResource.com Tutorial.

Eurocontrol. 2003, *The Development of Situation Awareness Measures in ATM Systems*. European Organisation for the Safety of Air Navigation. Retrieved June 18, 2004, at www.eurocontrol.int/humanfactors/docs/HF35-HRS-HSP-005-REP-01withsig.pdf.

Federal Aviation Administration. 1990, *Profile of Operational Errors in the National Aerospace System*, Technical Report. Washington, DC.

Fiore, S. M. Cuevas, H. M., and Oser, R. L. 2003, A Picture Is Worth a Thousand Connections: The Facilitative Effects of Diagrams on Mental Model Development, *Computers in Human Behavior*, Vol. 19, pp. 185–199.

Flach, J. M. 1995, Situation Awareness: Proceed with Caution, *Human Factors*, Vol. 37, No. 1, pp. 149–157.

Frantz, J. P. and Rhoades, T. P. 1993, A Task-Analytic Approach to the Temporal and Spatial Placement of Product Warnings, *Human Factors*, Vol. 35, pp. 719–730.

Giggerenzer, G. and Todd, P. 1999, *Simple Heuristics That Make Us Smart*. Oxford, UK: Oxford University Press.

Gordon, S. E. 1994, *Systematic Training Program Design*. Upper Saddle River, NJ: Prentice Hall.

Gray, W. D. and Fu, W. T. 2004, Soft Constraints in Interactive Behavior: The Case of Ignoring Perfect Knowledge in-the-World for Imperfect Knowledge in-the-Head, *Cognitive Science*, Vol. 28, pp. 359–382.

Hancock, P. A., Wulf, G., Thom, D., and Fassnacht, P. 1990, Driver Workload During Differing Driving Maneuvers, *Accident Analysis and Prevention*, Vol. 22, No. 3, pp. 281–290.

Hansen, J. P. 1995, An Experimental Investigation of Configural, Digital, and Temporal Information on Process Displays, *Human Factors*, Vol. 37, No. 3, pp. 539–552.

Hebb, D. O. 1955, *The Organization of Behavior*. New York, NY: John Wiley and Sons.

Hebb, D. O. 1976, Physiological Learning Theory, *Journal of Abnormal Child Psychology*, Vol. 4, No. 4, pp. 309–314.

Jedetski, J., Adelman, L., and Yeo, C. 2002, How Web Site Decision Technology Affects Consumers, *IEEE Internet Computing*, Vol. 6, No. 2, pp. 72–79.

Johnson, A. 2003, Procedural Memory and Skill Acquisition, in Healy, A. F., Proctor, R. W., and Weiner, I. B. (Eds.), *Handbook of Psychology, Experimental Psychology*. New York, NY: John Wiley.

Jones, M. and Polk, T. A. 2002, An Attractor Network Model of Serial Recall, *Cognitive Systems Research*, Vol. 3, pp. 45–55.

Katz, M. A. and Byrne, M. D. 2003, Effects of Scent and Breadth on Use of Site-Specific Search on e-Commerce Web Sites, *ACM Transactions on Computer-Human Interaction*, Vol. 10, No. 3, pp. 198–220.

Kemp, T. 2001, CRM stumbles amid usability shortcomings. *Internet Week Online*. April 6. Retrieved August 4, 2004, at www.internetweek.com/newslead01/lead040601.htm.

Khattak, A., Kanafani, A., and Le Colletter, E. 1994, Stated and Reported Route Diversion Behavior: Implications of Benefits of Advanced Traveler Information Systems, in *Transportation Research Record*, No. 1464, 28.

Klein, G. 2000, *Sources of Power*. Cambridge, MA: MIT Press.

Klein, G. A. 1993, A Recognition-Primed Decision (RPD) Model of Rapid Decision Making, in Klein, G. A., Orasanu, J., Calderwood, J., and MacGregor, D. (Eds.), *Decision Making in Action: Models and Methods*. Norwood, NJ: Ablex Publishing.

Konz, S. and Johnson, S. 2000, *Work Design: Industrial Ergonomics*. Scottsdale, AZ: Holcomb Hathaway.

Lachman, R., Lachman, J. L., and Butterfield, E. C. 1979, *Cognitive Psychology and Information Processing. Chapter 9 Semantic Memory*. New York, NY: John Wiley and Sons.

Lamoreux, T. 1997, The Influence of Aircraft Proximity Data on the Subjective Mental Workload of Controllers on the Air Traffic Control Task, *Ergonomics*, Vol. 42, No. 11, pp. 1482–1491.

Laughery, K. R. and Hammond, A. 1999, Overview, in Wogalter, M. S., DeJoy, D. M., and Laughery, K. R. (Eds.), *Warnings and Risk Communication*. Boca Raton, FL: Taylor and Francis Group.

Marble, J. L., Medema, H. D., and Hill, S. G. 2002, Examining Decision-Making Strategies Based on Information Acquisition and Information Search Time, in *Proceedings of the Human Factors and Ergonomics Society 46th Annual Meeting*, Human Factors and Ergonomics Society, Santa Monica, CA.

Miller, G. A. 1956, The Magical Number Seven, Plus or Minus Two, *Psychological Review*, Vol. 63, pp. 81–97.

Mullen, J. 2004, Investigating Factors That Influence Individual Safety Behavior at Work, *Journal of Safety Research*, Vol. 35, pp. 275–285.

Nielsen, J. 1999. When Bad Designs Become the Standard. *Alertbox*, November 14. Retrieved August 4, 2004, at www.useit.com/alertbox/991114.html.

Norman, D. A. 1988, *The Design of Everyday Things*. New York, NY: Basic Books.

Ordonez, L. and Benson, L. 1997, Decisions Under Time Pressure: How Time Constraint Affects Risky Decision Making, *Organizational Behavior and Human Performance*, Vol. 71, No. 2, pp. 121–140.

Overbye, T. J., Sun, Y., Wiegmann, D. A., and Rich, A. M. 2002, Human Factors Aspects of Power Systems Visualizations: An Empirical Investigation, *Electric Power Components and Systems*, Vol. 30, pp. 877–888.

Persaud, C. H. and Resnick, M. L. 2001, The Usability of Intelligent Vehicle Information Systems with Small Screen Interfaces, in *Proceedings of the Industrial Engineering and Management Systems Conference*, Institute of Industrial Engineers. Dallas, Texas, May, 21–23, 2001.

Rasmussen, J. 1993, Deciding and Doing: Decision Making in Natural Contexts, in Klein, G. A., Orasanu, J., Calderwood, J., and MacGregor, D. (Eds.), *Decision Making in Action: Models and Methods*. Norwood NJ: Ablex Publishing.

Reason, J. 1990, *Human Error*. New York, NY: Cambridge University Press.

Resnick, M. L. 2001, Task Based Evaluation in Error Analysis and Accident Prevention, in *Proceedings of the Human Factors and Ergonomics Society 45th Annual Conference*, Human Factors and Ergonomics Society. Minneapolis, Minnesota, October 8–12, 2001.

Resnick, M. L. 2003, Building the Executive Dashboard, in *Proceedings of the Human Factors and Ergonomics Society 47th Annual Conference*, Human Factors and Ergonomics Society. Denver, Colorado, October 13–17, 2003.

Resnick, M. L. 2014, Human Factors, in Badiru, A. B. (Ed.), *Handbook of Industrial and Systems Engineering*. Boca Raton, FL: Taylor and Francis Group, pp. 431–453.

Rodgers, M. D., Mogford, R. H., and Strauch B., 2000, Post Hoc Assessment of Situation Awareness in Air Traffic Control Incidents and Major Aircraft Accidents, in Endsley, M. R. and Garland, D. J. (Eds.), *Situation Awareness Analysis and Measurement*. Mahwah, NJ: Lawrence Erlbaum.

Roth, E. M., Bennett, K. B., and Woods, D. D. 1987, Human Interaction with an Intelligent Machine, *International Journal of Man-Machine Studies*, Vol. 27, pp. 479–525.

Roth, E. M., Gualtieri, J. W., Elm, W. C., and Potter, S. S. 2002, Scenario Development for Decision Support System Evaluation, in *Proceedings of the Human Factors and Ergonomics Society 46th Annual Meeting*, Human Factors and Ergonomics Society, Santa Monica, CA.

Sanders M. S. and McCormick E. J. 1993, *Human Factors in Engineering and Design*, 7th ed. New York, NY: McGraw-Hill.

Schwartz, B. 2004, *The Paradox of Choice*. New York, NY: HarperCollins.

Scielzo, S., Fiore, S. M., Cuevas, H. M., and Salas, E. 2002, The Utility of Mental Model Assessment in Diagnosing Cognitive and Metacognitive Processes for Complex Training, in *Proceedings of the Human Factors and Ergonomics Society 46th Annual Meeting*, Human Factors and Ergonomics Society, Santa Monica, CA.

Simon, H. A. 1955, A Behavioral Model of Rational Choice, *Quarterly Journal of Economics*, Vol. 69, pp. 99–118.

Snodgrass, J. G., Levy-Berger, G., and Haydon, M. 1985, *Human Experimental Psychology*. New York, NY: Oxford University Press.

Stevens, S. S. 1975, *Psychophysics*. New York, NY: Wiley.

Swain, A. D. and Guttman, H. E. 1983, *A Handbook of Human Reliability Analysis with Emphasis on Nuclear Power Plant Applications*. NUREG/CR-1278, Washington, DC: USNRC.

Swezey, R. W. and Llaneras, R. E. 1997, Models in Training and Instruction, in Salvendy G. (Ed.), *Handbook of Human Factors and Ergonomics*, 2nd ed. New York, NY: Wiley.

Taatgen, N. A. and Lee, F. J. 2003, Production Compilation: A Simple Mechanism to Model Complex Skill Acquisition, *Human Factors*, Vol. 45, No. 1, pp. 61–76.

Tulving, E. 1989, Remembering and Knowing the Past, *American Scientist*, Vol. 77, pp. 361–367.

Wickens, C. D. and Hollands, J. G. 2000, *Engineering Psychology and Human Performance*. Upper Saddle River, NJ: Prentice Hall.

Wickens, C. D., Lee, J. D., Liu, Y., and Gordon Becker, S. E. 2004, *An Introduction to Human Factors Engineering*, 2nd ed. Upper Saddle River, NJ: Prentice Hall.

Wogalter, M. S. and Leonard, S. D. 1999, Attention Capture and Maintenance, in Wogalter, M. S., DeJoy, D. M., and Laughery, K. R. (Eds.), *Warnings and Risk Communication*. Boca Raton, FL: Taylor and Francis Group.

CHAPTER 10

Industrial Engineering and Digital Engineering

I NDUSTRIAL ENGINEERING PROVIDES A linkage between traditional manufacturing and the emergence of additive manufacturing (AM), which is direct digital manufacturing (DDM). There are several additive manufacturing activities going on in various defense-related industries around the world. Of particular interest are the leading-edge initiatives going on at the U.S. Air Force Research Laboratory (AFRL) and the U.S. Air Force Institute of Technology (AFIT). The AFIT work and laboratory development were facilitated under the leadership and vision of an industrial engineer, yours truly (Badiru et al., 2017). Thus, the typical efficiency, effectiveness, and productivity principles of industrial engineering were applied to the AM development at AFIT.

AFRL, located within the Wright-Patterson Air Force Base (WPAFB) in Dayton, Ohio, is a premier research facility of the U.S. Air Force. It is a global technical enterprise, boasting some of the best and brightest researchers and leaders in the world. The lab prides itself on being revolutionary, relevant, and responsive to the Warfighter and the nation's defense. It delivers its mission by unleashing the full power of scientific and technical innovation. This mission includes leading the discovery, development, and integration of affordable warfighting technologies for the nation's air, space, and cyberspace force. Additive manufacturing features prominently in the new innovation pursuits of AFRL. It is important to note that the city of Dayton is internationally recognized as the birthplace of aviation, thus demonstrating the city's heritage of innovation. The city's prestige of innovation continues today. Incorporating additive manufacturing into the city's portfolio of innovation fits the theme of this handbook. AFIT, colocated with AFRL at WPAFB, is an internationally recognized leader for defense-focused technical graduate and continuing education, research, and consultation. The graduate degrees offered by AFIT are predicated on thesis and dissertation research. It is through this research and development avenues that new additive manufacturing initiatives are being pursued at AFIT.

In 2017, AFIT made a significant investment in new state-of-the-art additive manufacturing equipment by purchasing a Concept Laser M2 Cusing sintering system. This equipment is globally seen as one of the most-desired high-end powder-bed-based laser metal additive manufacturing systems. Sintering, which is the use of pressure and heat below the melting point to bond metal particles, is the ultimate application of additive manufacturing. It is the metal-based printing rather than polymer-based printing of three-dimensional (3D) parts. With this equipment, part sizes can range from very tiny to extremely large, thereby creating opportunities to build a variety of parts meeting the needs of the defense industry. In LaserCUSING machines, application-specific 3D parts with enhanced performance profiles are created in a fully automated digital process. This will facilitate new research and development partnership opportunities between AFIT and collaborators in terms of cost, efficiency, effectiveness, relevance, flexibility, adaptability, modularity, and responsiveness of 3D-printed products. For benchmarking purposes, as of January 2017, Concept Laser's X line 2000R system is the largest metal sintering machine available on the market. AFIT is proud to invest in Concept-Laser equipment to facilitate research, instruction, and consultation on additive manufacturing.

On the AFRL side, a recent article in the WPAFB *Skywrighter* newspaper provides a good account of the latest additive manufacturing research and applications at the lab. For scientists and engineers at the Air Force Research Laboratory's Materials and Manufacturing Directorate, additive manufacturing, also known as 3D printing, can be a powerful tool for rapid innovation.

Ultimately, it is a new way of looking at manufacturing across the materials spectrum and an area with challenges and opportunities that the Air Force is meticulously exploring. "Additive manufacturing is a huge opportunity for us," said Jonathan Miller, a materials scientist and the additive manufacturing lead for the directorate. "It allows us to manufacture unique form factors; it provides the opportunity to add functionality and capability to structures that already exist. Essentially, it allows us to redefine manufacturing." Traditional manufacturing methods developed during the times of the Industrial Revolution, when machines began to overtake the human hand for mass production. Many of these processes required material to be molded or milled away from a larger form to produce a specific design. Additive manufacturing, by contrast, is defined by ASTM International as the process of joining materials together, layer by layer, based on 3D model data. It increases design possibilities, enhances the speed of innovation, and offers an alternative for creating shapes closer to what an engineer might need, with fewer constraints. "The biggest problem with conventional manufacturing processes is time," said Miller. "Manufacturing is an iterative process, and you never get a part 'just right' on the first try. You spend time creating the tools to manufacture a complex part and then spend more time when you realize an initial design needs to be modified.

Additive manufacturing offers lower cost tooling and lower lead times. The early mistakes don't hurt you as badly."

EARLY DAYS

Though additive manufacturing is receiving a lot of industry interest as of late, it is not new to AFRL. Research into this manufacturing capability for the Air Force started at the same time the concept of rapid prototyping emerged in industry in the 1980s. Rapid prototyping was based on the premise that if engineers had an idea and wanted to make a shape, they could visit a shop and "print" the object, usually out of plastic by a printer. "The focus at this time was on creating functional prototypes, or objects that resembled a desired part, but the materials lacked the strength for even minimal use," said Miller. Early additive processing used light to chemically react to specific regions in a volume of gel to build rigid, plastic parts. The technology further evolved to include fused filament modeling, wherein fibers of plastic thread were melted and joined together to form a new object. Additional powder-based processes made use of plastic flakes that were melted by a laser into a shape. In the early 1990s, scientists learned that similar additive manufacturing processes could be used for generating metal objects. However, the technology at the time resulted in crude, large parts with poor surfaces. It was not until the late 2000s that laser technology matured sufficiently to truly move forward in this domain. "This spurred the additive revolution pursued today by the entire aerospace industry," said Miller.

SHIFT TO PRODUCTION PARTS

While more affordable lasers and metal powder processes were helping scientists to make better metal products, the "glue gun" route to additive manufacturing of plastics became much cheaper. Small, inexpensive 3D printing machines began to turn up in garages and schools, to the amateur engineer's delight. "Collectively, these became a new way of thinking about how to make stuff," said Miller. As additive manufacturing thinking evolved from being a way to develop prototypes to a method for actual production, the benefits and applications for the Air Force grew enormously, along with the potential for it to do even more. The manufacturing of customized parts and unique, complex geometric shapes at low production quantities can help to maintain an aging aircraft fleet. Custom tools, engine components, and lightweight parts can enable better maintenance and aircraft longevity. "Additive manufacturing can address a multitude of challenges for us, and there is a big pull to implement these processes from the logistics community," said Miller. "The fleet is aging, and replacement parts for planes built 30 years ago often no longer exist. Rapid production of a small number of hard-to-find parts is extremely valuable." However, the need to develop consistent, quality materials for additive manufacturing still remains a challenge that AFRL researchers are working diligently to address. Engineers need to have full confidence in additive manufactured part alternatives as they implement them as replacements in aging fleets or as system-level enablers in new weapon systems. "There are limits as to how the Air Force can use this technology and for what applications it will work best," said Miller. "That research is the basis of our work here."

EXTENSION TO FUNCTIONAL APPLICATIONS

As additive manufacturing has matured over the past few decades, the field has broadened beyond plastic and metal parts. Dan Berrigan, the additive lead for functional materials at the directorate, is exploring ways to use additive manufacturing processes to embed functionality into structure, such as by adding electronic circuitry or antennas on nontraditional surfaces. As the demand for flexible devices such as activity trackers and performance monitors increases, so does a need to power these sources organically. "Additive processes enable us to deposit electronic devices in arbitrary shapes or in flexible, soft form factors," said Berrigan. "We are looking at different ways to make a circuit that can enable them to bend or adhere to new surfaces or geometries, such as on a dome or patch. Essentially, we are looking at ways to add capabilities to surfaces that already exist." Conventional circuit fabrication requires the lamination of a series of conductive and insulating layers in a patterned fashion, resulting in a rigid circuit board. The electronic properties for these circuits are known and understood, and engineers are able to ensure that the circuit can conduct as intended based on these known concepts. For 3D printed electronics, a conductive material is divided into millions of small pieces and suspended in a liquid that is then dispensed from a printer, explained Berrigan. After printing, those individual conductive pieces must maintain contact to enable electrons to move through a circuit and create power. "The demand here is for low-cost, flexible electronic devices, and these direct write, additive processes give the community design capabilities that we cannot achieve otherwise," said Berrigan.

ADDITIVE CHALLENGES AND FUTURE POTENTIAL

Despite years of development and research into additive manufacturing processes, there are a number of implementation challenges that AFRL researchers need to address in order to enable greater Air Force benefit from the technology, both now and in the future. The techniques of industrial engineering can help mitigate the present and future challenges. "Fundamentally, it comes down to a materials processing problem," said Berrigan. The lack of standardized production processes, quality assurance methods, significant material variability, and reduced material performance are just some of the factors AFRL researchers need to overcome. Depending on the application, material performance can be related to the strength of a part. For example, the electronic properties of an additive manufactured circuit may be worse than those of ones traditionally manufactured. "Understanding the safety, reliability and durability of a part is critical for an aircraft. We know this for parts made through other processes, but we don't know this yet for additive," said Berrigan. Another issue centers on basic materials compatibility. "There are a lot of different interfaces in additive manufacturing, and ensuring that materials adhere to one another or that a part can support a certain stress or withstand a certain temperature—these are all challenges we need to address," said Miller. The long-term goal, according to Berrigan, is for additive manufacturing to become a well-understood tool in an engineer's toolbox, so that unique components can be design-integrated into a system. It is difficult to go back in a system already built, he said, but additive manufacturing provides the opportunities to build-in greater potential at the start.

"The long-term vision is to have functional and structural additive manufacturing to work more cohesively from the start. Rethinking systems-level design to incorporate functionality such as electrical wiring, sensors or antennas is a potential that additive can help us address," he said. "When you build something by layer, why not introduce channels for sensors, cooling or other functions?" In all, AFRL researchers agree that continued research and time will lead to fuller implementation of additive processes for the Air Force systems of today and the future. Innovative technologies are enabling capabilities, and additive technology is one with limitless opportunities to explore.

SUMMARY

Additive manufacturing is revolutionizing manufacturing processes. Although traditional manufacturing has undergone major advancements and new technology developments in recent years, new opportunities are needed to further the advancement. Traditional manufacturing relies on tools and techniques developed over several decades of making products. Additive manufacturing, popularly known as 3D printing, brings the efforts to a new level of possibilities. These possibilities are dependent on research and development efforts by organizations such as the Air Force Research Lab and the Air Force Institute of Technology. Collaboration between these two organizations and others, such as the Maker-Movement facilities, will ensure that the much-touted benefits of this new tool continue to be realized far into the future.

On a historical note, the ICOM (inputs, controls, outputs, mechanisms) technique, which originated from the IDEF0 (Integrated DEFinition for business process modeling) methodology came out of AFRL research in manufacturing in the 1970s and 1980s. The IDEF acronym was originally the ICAM (Integrated Computer Aided Manufacturing) Definition language. The IDEF2 part of the project involved Alan Pritzker and had a huge influence on what would become SLAM, which was the simulation language of choice in the 1980s and 1990s.

Interestingly, I have my own academic linkage to Alan Pritzker, who was the Ph.D. advisor of Gary Whitehouse, who was my own Ph.D. advisor. Such a genealogical relationship is essential for the continuing advancement of industrial engineering in the present-day context. The AFRL Manufacturing Directorate originally funded the development of IDEF0 (activity models), IDEF1 (data models), IDEF2 (dynamic models), and IDEF3 (process models) with extensions up to IDEF10 developed out of Texas A&M University and a small company called KBSI and funded via the Air Force Human Resources Lab, now AFRL/RH.

CONCLUSION OF THE STORY

Based on the aforementioned intellectual and educational linkages, it is envisioned that industrial engineering will continue to play a role and have a place in future developments of digital engineering and other developments in business, industry, government, and the military. This is how the profession of industrial engineering will continue to be advanced, sustained, and spread globally.

REFERENCE

Badiru, A.B., Valencia, V.C. and Liu, D. 2017, *Additive Manufacturing Handbook: Product Development for the Defense Industry.* Boca Raton, FL: Taylor and Francis Group.

CHAPTER 11

Industrial Engineering Around and About

INTRODUCTION

Industrial engineering here and there.

Industrial engineering everywhere.

Everywhere industrial engineering.

Industrial engineering is a cornerstone of the practice of engineering. Engineering is the foundation for national development. Engineers work at the intersection of science, technology, and societal needs. Throughout history, engineering has played a crucial role in the advancement of commerce, development of society, and pursuit of human welfare. The application of engineering to the problems of society is predicated on structured education programs. This chapter addresses how industrial engineering education is progressively important to society and what must be done to continually advance the quality and effectiveness of industrial engineering education. The chapter proposes 15 grand challenges for global industrial engineering education, based on the premise of the 14 Grand Challenges for Engineering published in 2008 by the U.S. National Academy of Engineering (NAE). The chapter highlights the role of diverse disciplinary viewpoints needed to ensure that industrial engineering education addresses economic, cultural, and social factors that impinge on engineering solutions to societal problems. The premise of this chapter is to spark more interest in research into models, techniques, and tools essential for making industrial engineering education more robust in solving global society problems.

Due to the breadth and depth of the profession, the story of industrial engineering must be told from a multitude of angles. It is envisioned that this focus book will spark additional story-themed writings by other industrial engineering scholars and practitioners.

As has been seen in the preceding chapters, industrial engineering is up, around, and about everything progressive in business, industry, government, and the military. Through industrial engineering, we can transform an opaque business world into a transparent world. It is through the tools, techniques, and principles of industrial engineering that we can achieve vertical and horizontal integration in all processes. The industrial engineering profession is so much needed today, tomorrow, and in the future. We must entrench ourselves in efforts to educate and train the future industrial engineering workforce to address the diverse problems of the world. Personally, I do not just see industrial engineering as my profession. I see it as a much-needed panacea for a lot of ills in many operational settings.

In an article I wrote for the *Industrial Engineering* magazine Badiru and Baxi (1994), I opined that industrial engineering was quickly losing its identity as a focused profession due to the increasing fragmentation of the profession into ever-smaller specializations that take the profession from its core focus. I still see that trend today. For example, college graduates today tend to associate more with focused professional societies rather than the one main professional body for industrial engineers, the Institute of Industrial Engineering (IIE). This is a disturbing drift that may destroy the identity of the profession as we know it. The shifting name change of IIE as a professional body does not help the matter. Given the option, graduates will align with other specializations rather than industrial engineering. I have written about many approaches and strategies that can help avert the downward trend. A good example is Badiru (2015), the contents of which are reiterated in this chapter. The 1994 and 2015 articles are not alone. In 1962, a special issue of the *Journal of Industrial Engineering* (Volume 13, Number 5) edited by Cecil G. Johnson addressed current trends and philosophies in industrial engineering education. Unfortunately, we have missed or forgotten the lessons presented in that special issue. Even the *Journal of Industrial Engineering* has transitioned into oblivion and was replaced by successive publications with ever-changing names. The seminal *Journal of Industrial Engineering* could have been developed into a lasting anchor for immortalizing the name of the profession. The shifting positions have decimated the professional name over the years. Many of the industrial engineering academic departments covered in that 1962 issue of the journal no longer have stand-alone or identifiable industrial engineering academic departments. My own alma mater phased out industrial engineering as an academic major in 2010.

My advocacy is to continue to explore other avenues for applications of industrial engineering to sustain the profession both in name and in practice. One good potential for industrial engineering is the diverse application potential in NAE's 14 Grand Challenges for Engineering. The 14 Grand Challenges for Engineering compiled by NAE in 2008 have implications for the future of general engineering education and practice. Engineers, particularly industrial engineers, of the future will need diverse skills to tackle the multitude of issues and factors involved in adequately and successfully addressing the challenges. An extract from the NAE document on the 14 grand challenges reads as follows:

> In sum, governmental and institutional, political and economic, and personal and social barriers will repeatedly arise to impede the pursuit of solutions to problems. As they have throughout history, engineers will have to integrate their methods and solutions with the goals and desires of all society's members.

This statement emphasizes the relevance of a holistic systems thinking approach in solving the multifaceted global problems that we face now and will face in the future. That is, indeed, the professional territory of industrial engineering. Global situational awareness (Badiru, 2009) is essential for solving the world's most pressing problems. Robust industrial engineering education programs around the world will be the cornerstone for integrating multiple skills across geographical boundaries as well as across cultural divides. It is a systems world. Industrial engineers are systems thinkers. Whatever affects one subsystem of the global infrastructure will eventually percolate through the whole system. Industrial engineering, by virtue of its versatility and systems viewpoint, can be the anchor for solving the challenges. The 14 NAE Grand Challenges for Engineering can help an educator to target research and education directions to collectively solve those problems that affect the global society. The challenges highlight the relevance of a holistic systems thinking approach in solving the multifaceted global problems that we face now and will face in the future. This is in perfect alignment with the premise of the 15 grand challenges for engineering education proposed in this chapter. The 14 grand challenges are

1. *Make solar energy economical*

2. *Provide energy from fusion*

3. *Develop carbon sequestration methods*

4. *Manage the nitrogen cycle*

5. *Provide access to clean water*

6. *Restore and improve urban infrastructure*

7. *Advance health informatics*

8. *Engineer better medicines*

9. *Reverse-engineer the brain*

10. *Prevent nuclear terror*

11. *Secure cyberspace*

12. *Enhance virtual reality*

13. *Advance personalized learning*

14. *Engineer the tools of scientific discovery*

With the positive outcomes of these projects achieved, we can improve the quality of life for everyone, and our entire world can benefit positively. Most of the grand challenges focus on high-tech developmental issues. Yet, the problems of the world are best solved through integrated approaches focusing on social, cultural, political, and high-tech issues, which fit the practice of industrial engineering.

Using the tools of engineering requirements analysis, engineering practitioners must appreciate the underlying relationships among the elements of a problem, thereby ensuring that all factors are considered in a global holistic solution. The V-model of systems engineering is used by industrial engineers to link and solve complex problems. It is a graphical representation of a system's life cycle. The model, which relates solution techniques to problem scenarios, is directly applicable to semantic network of the grand challenges. Problem assessment, problem decomposition, life cycle requirements, solution integration, and global implementation are key elements of solving the 14 grand challenges. This is the core of the research and practice of industrial engineering.

Education is the avenue through which the goals and objectives of the 14 grand challenges can be realized. It could be education in terms of preparing the future engineers or education in the form of raising the awareness of members of the society. Energy, in its various forms, appears to be a common theme in many of the 14 grand challenges. In addition to teaching the technical and analytical topics related to energy, energy-requirement analysis must consider the social and cultural aspects involving the three primary focus areas as follows:

- *Energy Generation*: Fossil, Fission, Fusion, Renewable, Nonrenewable, Safety, Security, Energy Independence

- *Energy Distribution*: Transmission Technology, Hydrogen, Distributed Energy Sources, Market

- *Energy Consumption*: Transportation, Storage, Product Requirements, Conservation, Recovery, Recycling

These are all areas in which industrial engineering has excelled for decades. Related to the foregoing discussions, this section presents what I see as the 15 pressing and grand challenges for global engineering education. These are referred to as "Badiru's 15 Grand Challenges for Global Engineering Education."

1. Effect a systems view of the world in educational delivery modes and methods in order to leverage unique learning opportunities around the world.

2. Pursue integration and symbiosis of global academic programs.

   - Through global educational system integration, move toward a mutually assured advancement of engineering education.

   - Think global, but educate locally to fit domestic needs.

- Language diversity, for example, can expand thought and understanding to facilitate global communication, cooperation, and coordination.

3. Link engineering education to the present and future needs of society rather than use it just as a means to better employment.

4. Embrace all engineering disciplines in a collaborative, one-focus alliance toward addressing societal challenges.

5. Engage nonengineering disciplines, such as management and the humanities, in addressing high-value societal problems collectively.

- There are now medical humanities programs. Consider engineering humanities programs to put a human face to engineering solutions.

6. Adopt and adapt e-education to facilitate blended learning modes, flexibility of learning, and diversity of thought in a fast-paced society.

- Of interest in this regard is the evolution of measurement scales for pedagogy and andragogy.

7. Leverage social media tools and techniques to facilitate serious and rigorous transmission of knowledge.

8. Extend formal engineering education to encompass continuing engineering education and sustainability of learning.

9. Create a hybrid method of teaching what is researched and researching what is taught.

10. Inculcate global sensitivity into engineering education programs.

11. Include social responsibility in engineering education, research, and practice.

12. Make engineering solutions more human-centric solutions.

- Use engineering to solve real human problems. Keep engineering education relevant to the needs of society.

13. Teach representational modeling in engineering education.

- Modeling can provide historical connectivity to recognize the present as an output of the past and a pathway for the future.

14. Teach "Of-the-Moment Creativity" to spur innovation for the current, prevailing, and attendant problem.

15. Introduce engineering solution "ilities" covering feasibility, sustainability, viability, and desirability of engineering solution approaches.

CONCLUSION

Throughout history, engineering has answered the call of society to address specific challenges. With such answers comes a greater expectation of professional accountability.

Engineering education is the foundation for solving complex problems now and in the future. Industrial engineering is the practical side of the problem solutions. In order to educate and inspire present and future generations of engineering students to tackle the pressing challenges of the world, a global systems perspective must be pursued. The 14 Grand Challenges for Engineering highlight the pressing needs. But until we can come up with practical and viable models and templates, the challenges will remain only in terms of ideals rather than implementation ideas. Industrial engineering can and should help in moving concepts and ideas to the practical realm.

REFERENCES

Badiru, A. B. 2009, Global Situational Awareness Using Project Management, *Industrial Engineer*, Vol. 41, No. 11, pp. 22–26.

Badiru, A. B. 2012, Application of the DEJI Model for Aerospace Product Integration, *Journal of Aviation and Aerospace Perspectives (JAAP)*, Vol. 2, No. 2, pp. 20–34.

Badiru, A. B. 2014, Quality Insights: The DEJI Model for Quality Design, Evaluation, Justification, and Integration, *International Journal of Quality Engineering and Technology*, Vol. 4, No. 4, pp. 369–378.

Badiru, A. B. 2015, A Systems Model for Global Engineering Education: The 15 Grand Challenges, *Engineering Education Letters*, Vol. 1, No. 1, pp. 1–14.

Badiru, A. B. and Baxi, H. J. 1994, Industrial Engineering Education for the 21st Century, *Industrial Engineering*, Vol. 26, No. 7, pp. 66–68.

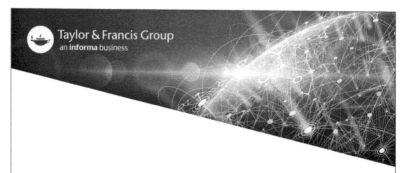